The Political Agency of British Migrants

This book offers a comparative analysis of the political agency of British migrants in Spain and France and explores how they struggle for a sense of belonging in the wake of Brexit.

With the UK's departure from the European Union (EU), Britons are set to lose EU citizenship as their political rights are redefined. This book examines the impacts this is having on Britons living in two EU countries. It moves beyond the political agency of underprivileged migrants to demonstrate that those who are relatively well-off also have political subjectivities: they can enter the political fray if their fundamental values or key interests are challenged. This book is based on ethnographic inquiry into the political agency of Britons in the Spanish Province of Alicante and South West France in the twenty-first century. Themes such as Britons becoming elected as local councillors in their countries of residence, migrants' reactions to Brexit, organisation of anti-Brexit campaigners, and claims for residency and citizenship are examined. The book foregrounds the contemporary practice theory built on the work of Pierre Bourdieu, as well as Engin Isin's approach to enacting citizenship, to provide empirical insights into the political participation of Britons. It does so by demonstrating how the elected councillors stood against gross moral inequity and fought for a sense of local belonging; how campaigners emoted digitally in reaction to Brexit; and how some migrants, keen to remain without worry, learnt both to navigate and to contest the policy and practice of national bureaucracies.

This book makes a first-ever contribution to the fields of anthropology and geography in the study of impacts of Brexit on British migrants within Europe. It is also the first study into lifestyle migrants as political agents. It will thus appeal to anthropologists, human geographers, sociologists, as well as academics and students of citizenship studies, migration studies, European studies, and political geography.

Fiona Ferbrache is Lecturer of Human Geography at Keble College and Brasenose College, University of Oxford. She is author of several publications on British migrants in France focussing on issues that include citizenship and Brexit. Fiona is also a transport geographer and has published two edited collections and several papers in that field.

Jeremy MacClancy is Professor of Anthropology, Oxford Brookes University. He has written and published extensively on the anthropologies of food, sport, art, and nationalism, as well as public anthropology and alternative histories of anthropology. He is at present researching the de- and re-population of 'emptied Spain'.

The Political Agency of British Migrants

Brexit and Belonging

Fiona Ferbrache and Jeremy MacClancy

Routledge
Taylor & Francis Group

LONDON AND NEW YORK

First published 2021
by Routledge
2 Park Square, Milton Park, Abingdon, Oxon OX14 4RN

and by Routledge
52 Vanderbilt Avenue, New York, NY 10017

Routledge is an imprint of the Taylor & Francis Group, an informa business

British Library Cataloguing-in-Publication Data
A catalogue record for this book is available from the British Library

Library of Congress Cataloging-in-Publication Data
A catalog record for this book has been requested

ISBN: 978-0-367-46297-0 (hbk)
ISBN: 978-1-003-02797-3 (ebk)

Typeset in Times New Roman
by Apex CoVantage, LLC

To all those who gave up their time to talk with us,
With our gratitude

Contents

Figures

Tables

Acknowledgements

We are jointly grateful to: the Department of Social Sciences, Oxford Brookes University, where we both of us at the time worked, for a 2016 Research Development grant for our work into town-councillors; and to the ESRC, for a 2017 IAA grant (C0RYSL00-AE01.11), which funded the Elche and Perigueux meetings.

Jeremy is deeply grateful to the Department of Social Sciences, Oxford Brookes University, for a 2017 Department Research grant, and to Oxford Brookes University for a 2018 Research Excellence and Impact Development grant, for his work on attitudes to Brexit among migrants; to the ERASMUS and ERASMUS+ programmes, which enabled annual teaching visits to Alicantine universities, where he conferred with Javier Úllan and Antonio Nogues; to the British consular staff, Alicante, particularly Sarah-Jane Morris and Sara Munsterhjelm, Consul and Vice-Consul, for his repeated encounters with them; to all those he spoke with, who generously gave up their time; to Bob Pomfret, Design and Media Services, Oxford Brookes University, for generously adapting the Alicantine map to our needs; to Elizabeth Hallam, Irmak Karademir Hazir, and Mikko Kuisma for reading sections of what became Chapter 4; and to the audiences at seminars Jeremy gave in the University of East Anglia, the University of Manchester, and Oxford Brookes University.

Fiona expresses particular thanks to the School of Geography and Faculty of Social Science and Business at University of Plymouth for a studentship to support her doctoral research on Britons in France; to University of Oxford for a 2019 ESRC/IAA grant enabling further work in this field; and to Jamie Quinn, University of Plymouth for his generous cartography skills. Fiona is extremely grateful to those who gave their time so openly to share thoughts and experiences within the research itself, and to audiences among whom Fiona presented and received feedback on the research in England and France: notably Judi and John, Mark Wise, Richard Yarwood, and James Sidaway; and to her parents, Muriel and Eric.

We have enjoyed working together on this manuscript. Each of us thinks our discussions have benefited our respective sections.

The order of our surnames is alphabetical; it has no further significance.

1 An introductory tour of our topics

We study the political behaviour of advantaged migrants in their country of residence. We study this because other scholars have neglected it. Yet, as we argue and demonstrate throughout this book, the phenomenon is of increasing importance in today's Europe. Some academics have researched and done extensive fieldwork on what they term 'lifestyle migration', the movement of those who seek to improve their quality of life (e,g, O'Reilly 2000; Benson 2011; MacClancy 2015). Others have investigated varieties of intra-European Union (EU) migration from the migrant's perspective (Favell 2008; Recchi and Favell 2009; Brändle 2018). Many more have studied the political actions of disadvantaged migrants based in Europe (e.g. Kofman et al. 2000; McNevin 2006; Oliveri 2012; Peró 2014; Cantat et al. 2019). But, until very recently, almost none has investigated in-depth the political life of migrants who are well-to-do or relatively so. Hence this book: an ethnographic delve into the political agency of Britons who have migrated to live in Spain and France.

Jeremy has long spent time in rural Spain, Fiona in the French country-side. We were both initially interested in what we considered an unusual activity in our respective areas: British migrants getting elected as local councillors. Since a change in EU electoral regulations (implemented 1997 in Spain, 1998 in France), an increasing number of them have acted as participative, provocative agents in the social and political lives of their new places of residence, as they sought to forge a renovated sense of who they are, what they can do, and where. We thus see our work as an investigation into political subjectivities and political agency i.e. peoples' conceptions of themselves and of what they might change within their world: a lived reality of plural institutions, multiple regulations, and uneven implementation. In 2015, we chose to work together, to study this topic comparatively. As our initial project was drawing to a close, the United Kingdom (UK) held its EU membership referendum (Brexit referendum). In both our fieldsites, the effect of the result on the British and their sense of self was so great, and the organisation of many into activist anti-Brexit groups so rapid, that following and analysing their progress seemed an obvious continuation of our politically oriented project.

We are not just interested in political agency. Both of us noted that many of the migrants' grave concerns over Brexit hinged on rights and EU citizenship: they thought the result of the referendum endangered their rights and status and at worst would lead to their retraction and the need to leave the host country. Also, we were deeply aware that EU citizens from other member states had only been allowed to stand in Spanish and French local elections because the European Parliament had in 1992 stretched the idea of EU citizenship to include this novel extension of local franchise. In other words, in municipal elections questions of citizenship were at least implicit, and in the debates about Brexit they were explicit, upfront, and disputed. Moreover, in recent years within academia, citizenship studies have moved far beyond their initial enclosure within legal studies, transforming into a lively, expanding arena of interdisciplinary activity (e.g. Ho 2008; Staeheli et al. 2012; Lazar 2013; Yarwood 2014; Paz 2019; Kallio et al. 2020). For these various reasons, both political and scholarly, we chose as one of our research foci the various ideas of EU citizenship at play within our ethnographic field of study. We wanted to know: who was deploying which version of this notion, why, how, and to what effect?

Any academic account needs a theoretical frame. And if that is not stated and worked with, chances are the implicit theory is left unexamined, its exercise unscrutinised, and its potential unrealised. To avoid those errors, we strive to be explicit. We found contemporary practice theory, inspired by the work of Pierre Bourdieu, to be particularly useful when attempting, in our initial chapters, to account for the motivations and experiences of British migrants, as well as the ideas and actions of those who would be municipal councillors. Also, other social scientists who have already worked on lifestyle migration in Western Europe relied on this approach and so it should not be overlooked (e.g. Oliver and O'Reilly 2010; Benson and Osbaldiston 2014). Practice theory is a renowned approach, of power and scope which accommodates multiple scales and a certain degree of fluidity. But it is a theory that focusses on social reproduction rather than social emergence. Thus, later in our text, to better understand the aims and behaviours of anti-Brexit campaigners, we turn to Engin Isin's approach which focusses on activist subjects enacting performative citizenship. He places radical, creative change at the very centre of his mode of explanation, with subjects constituting themselves as agents through the imaginative acts that they enable. Practice theory, as its proponents point out (e.g. O'Reilly 2012), is not an exclusionary theory but rather an explicatory framework within which others nestle. Thus, we later deploy both neo-Bourdieuesque and Isinic approaches to comprehend migrants' strategies to secure contin-ued residence within the loose mesh of supranational regulation, its national implementations, and bureaucratic realities. Since every approach has its limits, in our closing assessment we try to pinpoint areas within which Isin's approach might be strengthened. Criticism is not denigration, but recogni-tion that his suggestive work is worthy of sustained attention.

Throughout our respective research, migrants were not coolly discussing citizenship and rights in the abstract, as isolated concepts subject to a finely focussed deconstructive analysis, more befitting of a detached, tenured academic. Rather, they were key concerns within a loose matrix of hotly debated ideas, ones central to the experiences of these migrants: besides citizenship and rights, that mesh includes residency, status, and belonging, the whole underpinned by a strong sense of morality. These contested con-cepts thus became further foci of our work, leading us to explore the ways migrants strove to control their own destinies, and to negotiate and fight against the categorical labyrinths created by nation-state bureaucracies.

For agitated migrants, the stakes were lofty, and their passions equally high, turning affective dimensions into structuring devices of their activi-ties. As interviewees in Spain stated, municipal (mis)management of their place of residence goaded some into reflecting on what kind of sociality was minimally acceptable, and then putting that into effect by standing for office. The referendum and its aftermath had even deeper, broader effect, marshalling migrants into different camps and creating new socialities. For Brexit in particular, people were articulate about the affective as a driver of political encounter. This is central for, as Cook et al. put it,

> If accounts of political dissatisfaction are to have either intellectual credibility or practical value, they have to be grounded in the specifici-ties of citizens' subjective experiences, with particular attention paid to the factors that structure their visions of how best to relate to them-selves and others, as well as to ideas, objects and institutions.
>
> (Cook et al. 2016:3)

We have therefore striven to extricate these affective engagements as much as possible throughout our account, evidencing the transformative power of different emotions, indigenously conceived, whether positive or negative, at generating actions, agglutinatory as well as divisive. In this way, our book is a critical foray into the affective creation and consequences of contempo-rary political subjectivities and actions, one which we hope blends well with discriminating use of practice theory and Isin's approach.

But, before we enter the ethnography proper, in the remainder of this introduction, let us clarify our approach to political agency in these settings, dwell briefly on EU citizenship, clarify our use of otherwise misleading terms, and then account for our field methods.[1]

Approaches to political activity

Some of the theoretically most sophisticated approaches to lifestyle migration have been inspired by Bourdieu's theory of practice (e.g. Bourdieu 1977; Bousiou 2008; Janoschka 2009, 2011; O'Reilly 2012; Lawson 2016). For these authors, Bourdieu's work is valuable because it can integrate micro- and macro-levels of analysis; attempts, via his concept of habitus, to transcend dualisms of structure and agency, materialism and idealism; and deploys a fine-grained approach towards the investigation of social class and its sub-sections. For example, Oliver and O'Reilly show how migrants might move in order 'to start a new life' only to find that non-economic forms of capital come to loom large and divide the incomers according to other resources (Oliver and O'Reilly 2010). Also, the 'clean' break with their past which many migrants claim to desire turns out in fact to be constrained by their habitus. It is not so much that they cannot realise their dreams; rather the constitution of their dreams is part of the past they wished to leave behind. 'In other words, their relative symbolic capital (incorporating educational, cultural, and social capital) impacts on the decision to migrate and the destinations chosen, but also the life then led in the destination' (Benson and O'Reilly 2009:618).

Bourdieu's approach can be criticised for its 'ethnocentric' conceptions of social class, elitist modes of distinction, and other universalising notions, while his focus on reproduction of social forms is claimed to leave little room for change (Jenkins 1992; Bousiou 2008:22; Beigel 2009:20–1; Gemperle 2009:14; Woodward and Emmison 2009:2; Daloz 2013). In response, Bourdieu claimed that he had intended habitus as a generative structure, albeit one whose generative capacity is limited by conditions of the time and place of its production (Bourdieu 1990). Thus, his later elaborated idea of habitus can be made to accommodate 'invention and improvisation, such as lifestyle projects' (Benson and O'Reilly 2009:617).

Contemporary practice theorists, building on Bourdieu's work while fully acknowledging its limits, seek to overcome some of the previous difficulties and broach more centrally the analysis of change. Mouzelis calls for the need to recognise both intra-habitus tensions and interaction between actors in a field (Mouzelis 2007). Kemp argues that habitus needs to be integrated with reflexivity, to provide scope for individual agency (Kemp 2010). O'Reilly, a leading exponent of modern practice theory, attempts to synthesise many of these corrective elements in a broad overarching approach to migration studies, which strives to take account of external structures, internal structures (including habitus), communities of practice, and outcomes. Of habitus she states that it is both 'fairly fixed and transposable' and 'constantly changing and adapting', though she does not detail the relation between fixture and change (O'Reilly 2012:151, 160). While her synthetic approach to practice theory is broadly encompassing, she recognises it 'merely provides the meta-theoretical frame within which disparate studies can be brought together. It does not attempt to do all the work that other theories and concepts contribute' (ibid.:84). She affirms her 'careful and critical' eclectic style of practice theory should 'not be applied too rigorously': it is, after all, a heuristic framework (ibid.:7, 160). Postill argues in a parallel manner that practice theory 'cannot be a theoretical cure-all'; e.g. it cannot tackle a 'world-historical moment', such as the Danish Muhammed cartoons controversy, caused by the September 2015 publication in a Danish newspaper of cartoons depicting Mohammed, which he considers more a political process than a social practice (Postill 2010:12–13). Comparable comments could be made about Brexit: it is above all an extended process, and as we go on to show, the anti-Brexit campaign is not a set practice but an innovative part of that process.

For Swartz, Bourdieu expected everyday disalignment between habitus and particular fields: mild disjuncture leads to adaptation, a gradual modification of structures; considerable disjuncture to transformation, producing resignation or revolt. Yet how these protests can lead to change is left unsaid: Bourdieu did not develop a politics of habitus (Swartz 2013:236–41). Swartz's concerns are bared in the illuminating work of modern practice theorist Michael Janoschka, who like Jeremy, worked in Alicante province, and on politically active migrants: the mid-2000s successful campaigners against mass-expropriation programmes in the Valencian Community (which includes Alicante); they were protesting against local politicians' abuse of legislation in order to urbanise private land and charge the landowners for doing so.[2] Janoschka recognises practice theory, 'developed in and exemplified by a virtually pre-modern society, . . . has certain shortcomings if the mobile conditions of people, capital, knowledge and practices in late modernity are reckoned' (Janoschka 2011:228; also Kemp 2010:156). However, 'a rapid and shock-like transformation' can 'produce a field of critical attitude that requires new interpretations and incorporations of the social world' (ibid.: 228). Thus, threatened incomers in Valencia established 'a temporarily radicalized habitus', on a par with Bourdieu's late invention of a 'subversive habitus' (ibid.:234; Bourdieu 2005). But Janoschka declares he cannot tell if the consciousness of the usual habitus dispositions endures and 'the weight of the reified world is still felt', or if a prolongation of the crisis turns the provisionally radicalised habitus into a permanently reconstituted one (ibid.:229). We suggest an empirical answer to his dilemma later in our text.

Janoschka's response to the challenges of his fieldsite is imaginative, but he makes habitus do too much work. He stretches it into a catch-all concept so malleable nothing can escape its reach. This version of praxis puts it on a logical par with Darwinianism: all examples, no matter how seemingly aberrant, can be made to fit into its theoretical schema. In sum, when powerful external forces, on the very margins of the quotidian habitus, stimulate a diversity of creative responses in people trying to imagine, and to influence their place in an open-ended future, then expanded versions of the concept of habitus are stretched to new levels of elasticity, with all the tensions that entails. The relative inertia of habitus, unless defined in O'Reilly's terms, is not easy to reconcile with the consideration of radical, lasting alteration. At the very least, just as a Darwinian framework forces one to think in terms of evolution, so deployment of habitus reminds us that dispositions are not wholly individual but structured, perhaps malleable but not completely open to fundamental change. Given this orientation of habitus towards more stable dispositions, it seems advisable, when studying fast-moving processes of significant change, such as Brexit, to look elsewhere, for instance to the work of Engin Isin on politics within Europe.

We are not antagonistic to contemporary practice theory. When it is pitched at such a high level as O'Reilly's informative approach, it is complementary to, not conflictive with alternative theories, e.g. Isin's. In this book, one question we ask is, which explanatory frameworks are the most suitable when accounting for gradual change or for abrupt alteration. Which is the more telling, for each context?

In his evolving work on citizenship in Europe, Isin is not concerned with social reproduction but with social emergence (Isin 2008). For that reason, he wishes to avoid the downsides of conventional modes of thought in the political sciences, which otherwise threaten to confine new configurations of social process within traditional taxa. He wants to catch, and follow the provisional, fluid nature of challenging processes in a vocabulary unfettered by the confining connotations of established terms, developed in times of different circumstances, when nation-states seemed sovereign and groups clearly defined. Isin's approach, therefore, is particularly relevant as a lens to examine emergent and unprecedented events such as Brexit.

To free himself from those tired vocabularies, Isin devises a novel nomenclature and procedures for studying what he terms 'acts of citizenship' (Isin 2013:21–8): 'the moment in which a subject – a person, a collective – asserts a right of entitlement to a liveable life when no such prior authorization exists, when no clearly enabling convention is in place' (Butler 2004 in Isin 2013:24). This is an activist-centred approach, which does not concentrate on rights as legal rules upheld by authoritative bodies integral to the nation-state or supranational body, i.e. the EU. Isin sees citizenship as much a bundle of legal rights as 'a social process through which individuals and social groups engage in claiming, expanding or losing rights' (Isin and Turner 2002:4). Thus, his emphasis tends towards the study of practices, meanings, and identities rather than legal rights and norms.

Rather than researching even relatively stable social states, Isin focusses on unfolding, future-oriented, exploratory, contested processes, which at the same time form its activist protagonists into groups. For him there is no fixed EU polity but rather a complex European juridico-political space, composed of elements and arrangements, i.e. the legal and constitutional foundations of citizenship, which are sometimes contradictory, sometimes complementary. Citizenship here is not a stable category but contingent, dynamic, and performed (Isin 2017). Subjects make claims to rights, which invoke and challenge the arrangements within the broad assemblage of European institutions. Claimants traverse different sites and scale of rights which may cross a range of these institutions, and may do so in an unorthodox manner which confronts 'dominant understandings of citizenship as membership in a contained polity' (Isin and Saward 2013:15). Coming together as common claimants, transcending boundaries as they go, generates new sites of belonging and identification: new subjectivities. Sites and scales are not pre-fixed but fluid, relational; they are formed through contest. Thus Isin's 'enactments of citizenship' are performative acts which produce subjects who collectively challenge present configurations, and which start to take what they ask for. Rights are not top-down exercises in abstract legalese but here take on a social reality because they are claimed, usually bottom-up (Ferbrache 2019a; MacClancy 2000b), by those whose performance of claim-seeking constitutes them as political actors and political subjects. This is a theoretical approach empirically focussed on change and the conflicts which go with it, which examines topics confined previously to the legal, in a variety of interconnected dimensions: political, ethical, cultural, sexual, and social (Isin 2013:41). In particular, it is a shift from conventional rights-based studies, grounded on black-letter law, to ones more focussed on processual, political, and performative dimensions.

Isin is explicit about the distance of his approach from Bourdieu and his followers. Practice theory attends to social formation, Isin to its reformation; practice theory focusses on subjects who conduct themselves, Isin on subjects who act (Isin 2012:108–9). He recognises practice theorists, to their credit, have demonstrated the role within citizenship of habitus 'formed over a relatively long period of time': but 'the question of how subjects become claimants . . . within a relatively short period of time has remained unexplored' (Isin 2008:17). For Isin, acts stand in contrast to habitus. Habitus emphasises relatively enduring dispositions and accounts for 'the persistence of an order'; acts are purposive, performative ruptures, which create disturbance. These ruptures enable actors (created by the acts) 'to create a scene rather than follow a script' (Isin 2009:379), and he distinguishes between 'active citizens' who follow scripted pathways instituted by governments, and 'activist citizens' who write their own. Thus, for Isin, here moving beyond practice theory, 'to be a citizen is to make claims to justice, to break habitus' (ibid.:384) in creative, unconventional, and unauthorised ways.

A further contrast between these two approaches is scale. Practice theory is all-encompassing in scope, seeking to knit meso- and micro-level events with macro-level social structures (e.g. globalisation), which individuals have relatively little ability to control or manipulate. We might call Isin's aims more demographically modest. He is not trying to explain whole societies. While including macro-level dimensions within the complex contexts of his localised case-studies, he concentrates on activist attacks to structure rather than the structures themselves, however enduring or adaptable they might be, i.e. if he pushes structure out of the foreground he replaces it with actors who enact.

Often, Isin argues, 'we see that subjects that are not citizens act as citizens: they constitute themselves as those with "the right to claim rights"' (Isin 2009:371). 'Acts of citizenship' have tended to focus on those who are not already citizens (e.g. Barbero 2012; McNevin 2012). Those with non-citizen status (e.g. refugees, irregular migrants, mobile populations) often exiting at the margins of (national) society, contest and construct citizenship into being as well as transform it. At the same time, 'insiders' holding status as national citizens can also express themselves through acts of citizenship in defence of what they take as their EU-grounded rights (López-Sala 2019). For instance, Britons on the Continent know they will remain EU citizens only until the details of the departure of the UK are formally agreed, when this status and associated rights will most likely be withdrawn. Though not yet non-EU citizens, they suffer 'precarious citizenship', i.e. they face the uncertainty of being able to secure permanent access to the rights which they enjoy as EU citizens (Lori 2017:3). This is comparable to 'liminal legality', where individuals exist in a 'grey area between undocumented and documented immigration categories' and lack 'the stability offered by permanent residency and citizenship' (Menjivar 2006; Birkvad 2019:801). This common degree of insecurity characterises both precarious citizenship and liminal legality. By examining Britons' (re)claims to rights, we seek to expand Isin's work into the realm of precarious citizens. But, first let us consider the citizenship status underpinning much of what this book is about.

From EU citizenship to precarious status

EU citizenship was legally established in 1992. It developed out of the postwar European project which united six European states towards the political goal of 'ever closer union' (Treaty of Rome 1957). In 1957, Belgium, France, Germany, Italy, Luxembourg, and the Netherlands became the founding members of the European Economic Community (EEC), forming the Single Market based on the free movement of capital, goods, services and labour. This internal market was progressively established via a series of treaties between the member states, expanding to become a union of 28 members (including the UK) and renamed the European Union in 1993 (Treaty of Maastricht 1993).[3] Its institutions have also grown and evolved to cover a wider range of political, economic, social, and cultural policy areas (Wise and Gibb 1993; Shore 2000). It is within the broadening of this political and economic project that EU citizenship emerged.

Introduced via the Maastricht Treaty and built upon the principle of freedom of movement,

> Citizenship of the Union confers on every citizen of the Union a primary and individual right to move and reside freely within the territory of the Member States, subject to the limitations and conditions laid down in the Treaty and to the measures adopted to give it effect.
>
> (Directive 2004/38/EC)

From a political perspective, free movement can be seen to promote European integration at the human level. For example, the Directive on Free Mobility (2004/38/EC) states:

> Enjoyment of permanent residence by Union citizens who have chosen to settle long term in the host member state would strengthen the feeling of Union citizenship and is a key element in promoting social cohesion, which is one of the fundamental objectives of the Union.

Political aspirations for 'an ever closer union' partially rest upon EU citizens who move to live, study, work, and retire outside of their national territories (Diez Medrano 2008; Recchi 2008; Favell and Recchi 2009). In the 1950s, the European economic bloc had already established free movement, but for certain types of valued workers and their families.[4] One novelty of EU citizenship was extension of free movement rights to all individuals based on nationality (of a member state) rather than employment. However, the ability to exercise these rights favours economically active migrants: that free movers are 'subject to the limitations and conditions' is a euphemism for the stipulation that any EU citizen wishing to reside in a host country must prove they have sufficient financial resources and sickness insurance not to become a burden on that country's social welfare system (some of the Britons we interviewed may have been relatively unaware of this until the Brexit referendum) (Seubert 2018). Furthermore, those wishing to stay more than three months in a member state are subject to various regulations within that country. For example, most member states require EU citizens to register. Registry regulations vary across the member states and some, such as France and the UK, have not had compulsory systems. This has led to uneven experiences of settling in France and Spain as we later demonstrate.

The European Commission has also implemented a range of regulatory policies to support not just free movement but equal treatment, of EU citizens alongside national citizens in member states (Recchi 2008; Seubert 2018). These policies include prohibiting discrimination on grounds of nationality, recognition of professional qualifications and educational degrees, and provision of European healthcare. Measures may help to lower personal, financial, social, and political costs of moving between European states. Although some of the Britons we interviewed migrated before EU citizenship was implemented, their daily lives became increasingly circumscribed by EU policies such as these. Moreover, these policies privilege EU citizen migrants in comparison with non-EU citizens entering the Union (see, for example van Houtum and Boedeltje 2009). However, even for EU citizens, free movement is far from frictionless (Diez Medrano 2008; Favell 2008), and the process and experiences of moving are extremely unequal (see, for example Guild 2016; Benson and Lewis 2019).

For the political theorist Sandra Seubert, EU citizenship offers opportunity: 'a chance of re-imagining and anticipating new forms of political agency and subjectivity' (Seubert 2018:2). Yet in practice, citizenship is often taken for granted, rarely questioned until it ceases to function as expected or is removed (Isin 2013; Ferbrache and Yarwood 2015). Pre-Brexit, Britons in northern France rarely acknowledged their EU citizenship status or the rights it bestowed, even when prompted to do so (Drake and Collard 2008). Similarly, their compatriots in South West France largely took EU citizenship for granted (Ferbrache and Yarwood 2015). The Brexit referendum of 23 June 2016 challenged this status quo, as 51.9% of those voting opted to leave the EU. This outcome foresaw British citizens losing EU citizenship; as we demonstrate, it also brought citizenship more centrally into the discourse and subjectivities of Britons in Spain and France.

The UK left the EU on 31 January 2020 and entered a transition period due to end on 31 December 2020. While there are many critical consequences of the UK's withdrawal, our focus here is on the removal of EU citizenship and its associated rights, given its centrality to the lives of the British migrants we research. Anticipated removal of EU citizenship leaves Britons in a tenuous legal position between their documented status as EU citizen and that of non-EU immigrant or third country national. Even as the UK entered transition, there was no certainty concerning the future rights of Britons living in other EU states (see Benson 2020; O'Reilly 2020). Birkvad argues that this type of 'liminal state can serve as a freeing and empowering social transformation, but if it extends indefinitely, it may produce enduring uncertainty and anxiety' (Birkvad 2019:801). As we demonstrate in this book, Brexit becomes a catalyst for both.

Misleading terms and numbers

A note on terminology: some of the people we study call themselves 'expatriates'. Others think this deeply inappropriate and think of convenient alternatives. Use of 'expatriate' may be questioned on several grounds. First, it has strong colonialist connotations, redolent of the imperial enterprise, as though it is more important where one comes from than where one currently is. Second, it may be viewed as potentially racist and gendered, especially when employment of 'expatriate' is restricted to West Europeans on the move, with 'migrant' as the default label for all others crossing borders (e.g. Remarque Koutonin 2015). For reasons such as these, the leading pressure group Bremain in Spain prefers to use 'migrant' over 'expatriate', 'British national resident in other EU country', or other clumsy multi-word phrase. Yet many Britons in Spain and France, including some anti-Brexit activists, are ready to use 'expatriate' without concern, while Alicantine sociologists have noted that, in the province, the 'collective imaginary' rejects 'immigrant' as a term for local residents from other EU member-states, as the majority do not seek to insert themselves into the labour market (Simó-Noguera et al. 2005:1). We recognise that all terms here have limitations and come with connotations. 'Migrant' seems to have the least so that is one term that we use in this book. Sometimes we shorten 'British migrants' to 'Britons' to avoid repetition.

We also use 'citizenship' to identify these migrants. 'British citizens' defines those who do not hold Spanish or French citizenship unless indicated otherwise in our discussion (the vast majority of those we interviewed). We find this definition useful to identify those most likely to be affected by the UK's decision to leave the EU. However, we recognise that 'British citizens', as with some of the other terms employed, is not an homogenous category. As indicated in the British Nationality Act 1981, not all those who are labelled 'British Citizen' hold the same rights. 'Channel Island British citizens', for example, are not EU citizens, and British citizens resident in the Channel Islands (whether EU citizen or not) have not been enfranchised to vote in UK elections, including the Brexit referendum.

We try also to avoid casual use of the term 'resident' in this book and use it distinctly to refer to the regulatory system of registration for migrants in these countries. Thus, it differs from the action of 'residing', which we prefer, and which serves to underline the continuing nature of the activity, over a claimed end-state: 'residence'.

Social class: both of us are well aware of the common stereotypes upheld by many in Britain, of the British in Spain as predominantly working-class or lower-middle class in origin, and of those in France as mainly upper-middle class in background. This did not square comfortably with our personal experience in our specific areas, nor with the range of our respective interviewees. We give further detail in later chapters on the diversity of people each of us interviewed. Social class was never mentioned by any of our interviewees, though some in France now spoke of having a more satisfying social life compared to the one they had enjoyed in the UK. Therefore, we did not consider the social class, if that could even be identified in an unambiguous manner, of those we each interviewed. In our fieldwork, it did not appear usually to be a relevant or revealing factor.

A word of caution: researchers of social arenas are often tempted to study the vaunted 'identity' of people(s) within their field area. This can be relatively easy to do, especially in situations where a change in circumstance makes people question or proclaim their own identities and values. For a fieldworker, this can make them a clear topic to focus on, amidst the otherwise seemingly liquid nature of so much contemporary life. The danger is that, particularly in these contexts, the identity may appear totalising because that is the desired end of its promoters. But totalising efforts are always contested, whether weakly or strongly, spoken or unvoiced (Cannadine 2013). Attempts to maintain some degrees of conviviality continue, despite the energetic efforts of flag-wavers (MacClancy 2016). The debate over Brexit is an obvious example of this two-faced process. The controversy has caused deep divide, great hurt. But, as we will demonstrate, many migrants, however strongly they held to their positions, strove to maintain a sense of even broader values.

Numerics: there is an array of sources which quote different numbers for the total population of Britons in Spain and in France. This is mainly a result of empirical and methodological difficulties of calculation. Partly, one is trying to count and estimate an often mobile population which is inherently difficult to track within the EU. In addition, there is a lack of systematic and official means of doing so in a sustained manner. The British Government, for example, does not maintain a record of emigrating citizens, nor does the French Government keep one on immigrants. Thus, the figures which appear in this book, though they may be drawn from the most reliable sources we can find, are usually not accurate but indicative.

Translations: in most cases we employ English words to convey meaning. Occasionally, however, some Spanish and French words are less easily translatable, or their English equivalent was seldom used by interviewees. In such cases, we maintain the original Spanish and French terms in the text. These include Spanish words such as *padrón* (municipal register) and *urbanización* (housing estate), as well as French ones such as *maire* (mayor), *mairie* (town-hall), and *commune* (the smallest French administrative division). Such terms are presented in italics only the first time they are employed in the subsequent narrative.

Methods

Jeremy is a social anthropologist, Fiona a human geographer. We both practise fieldwork, i.e. the intensive study of communities via participant-observation. Besides gaining insights from spending time with people and serendipitous encounter, we conduct semi-structured interviews with targeted individuals, usually as many as we can until we reach a high degree of semantic saturation. In other words, once successive interviews tell us very little new, we tend to stop seeking further candidates for interviewees.

Like most of our interviewees, we are also British citizens, though Fiona is a British citizen of the Channel Islands, which has various nuances and exclusions, and Jeremy, thanks to his father, bears an Irish passport as well as a UK one; his mother is of Maltese ancestry. Jeremy was awarded his doctorate in 1983, based on his fieldwork in Vanuatu; Fiona gained hers in 2011 (Ferbrache 2011b). We are both white with a base in our respective fieldsites. Jeremy and his wife have owned a rural secondary residence in the hills above Alicante since 2005. A fluent speaker of Spanish, thanks to his postdoctoral fieldwork in Navarre and the Basque Country (MacClancy 2000a, 2007), he has visited the area several times every year since then. His circle of acquaintance, both Spanish and British, includes neighbours, villagers, service-providers, and academics in both of the Alicantine universities, where he gives a week-long Erasmus teaching course annually. Fiona's research site in Occitanie is also where she lives. Initially arriving in 2008 to carry out doctoral research, and affiliated with Université de Toulouse-II-Le Mirail, she never truly left. Prolonged engagement with the area has generated data spanning more than ten years, while her personal experiences have facilitated deep insight and rapport.

Our respective fieldsites chose us as much as we chose them. We do not think that either one can be taken as representative of British migrant opinion on the Continent. More to the point, we cannot think of any one place which could: each has its own historical contingencies and local trajectories. However, intensive investigation in any one particular spot can reveal the key ideas and their interrelations held by those we wish to study. And comparison of parallel but different populations in contiguous countries may give us an idea of the particularity and generality of the processes we analyse. To put that another way, we are not producing a pair of authoritative portraits defining all their like but rather a detailed diptych suggesting a bigger picture.

The analysis in this book is based on extensive data from several periods of research. In Jeremy's case since 2016, and from 2008 in Fiona's. For the material on the British in Spain, Jeremy interviewed in 2016 and 2017 over 25 people, some several times: practising or former British councillors, their Spanish opponents, their compatriot acquaintance; British consular officials; Catherine Bearder, then the only Liberal Democrat Member of the European Parliament (MEP); British office-holders in local political and socially oriented organisations. In early 2018, the overseer of the Facebook page for the French-based, anti-Brexit campaign group FFU gave Jeremy permission to access its discussion group. In Spring 2019, he carried out a separate series of interviews with 24 British residents in the Alicantine area about their reactions to Brexit, plus their fears and their hopes for the future. In all cases of face-to-face encounters, Jeremy recruited interviewees by word-of-mouth or, to a much lesser extent, via public calls in relevant local media or online discussion groups.

Jeremy failed to interview sufficient members, for his satisfaction, of one particular sub-group of British migrants to Spain: the 'unregistered'. Their numbers in the country can only be guesstimated as they shy from bureaucracy, but they are generally thought to be high. Yet this 'hidden' population is relatively unknown for two kinds of reasons: besides being suspicious of anyone who smacks of officialdom, which may include fieldworkers, they are also under-researched for structural reasons of academic funding. Twice we sought funds to study members of this group specifically. We tried to assure research grant assessors of our already established acquaintance with some 'unregistered' UK migrants to no avail. We were informed we were not providing sufficient guarantees of access to this category of migrants. In effect, the contemporary criteria of British Government-backed funding into the social sciences are set to exclude those researchers who cannot provide almost risk-free evidence of entry into a target demographic whose members think they have reason to be wary of investigators. To our knowledge, the only systematic work so far carried out on the 'unregistered' was a face-to-face survey of the local population conducted voluntarily by a municipal political party, run predominantly by migrants, in southern Alicante: in other words, long-term residents interviewing their neighbours. Jeremy only managed to interview a few of the 'unregistered' thanks to longstanding friends who acted as intermediaries trusted by both parties.

Material on Britons in France comes from Fiona's 53 interviews (with individuals, couples, and families) and participant observation in 2008–09 in the former south west region of Midi-Pyrénées.[5] Fiona mainly interviewed Britons in rural areas to the east of Toulouse (the region's capital), but ten were based in the centre or peri-urban outskirts of the city. British activity groups (e.g. a walking group) and webpages (Angloinfo.com) were used to begin a snow-balling process of meeting and recruiting Britons for inter-views. Snow-balling gathered momentum through word-of-mouth contacts, and Fiona sought diversity within that. Between April and July 2016, Fiona interviewed a separate group of 15 former and practising British councillors, and a representative of the British Consulate in France. She used a gatekeeper to make initial contacts and then recruited through word-of-mouth and avail-ability sampling. In August 2016, Fiona carried out an online survey among 100 Britons residing in France to gauge their reactions to Brexit. The sur-vey explored emotions, concerns, and actions through four open questions. While the survey was distributed to known Britons and via online platforms used by Britons in France, it was not possible to verify that those responding were actually British migrants in the country. The findings indicated some initial patterns which were used in a subsequent set of 18 interviews examin-ing Britons' responses to Brexit.

Alongside planned interviews, we take advantage of our habitual moments in the field. Fiona's material is supplemented by numerous spontaneous interviews and encounters since 2008, Jeremy's since 2005. In contrast to recorded and transcribed formal interviews – where possible – fieldnotes are recorded in writing often where they take place (e.g. cafes, markets, events, grocery stores), otherwise as soon as possible afterwards. In addition, as scavenger-fieldworkers, we also take advantage of any further source of information we scour or come upon: flyers, posters, websites, online fora, official documents, as well as newspapers and magazines, whether produced regionally, nationally, or overseas.

Together, in Spring 2017 we staged meetings, funded by the Economic and Social Research Council (ESRC), UK, in Alicante and Perigueux on Brexit and British migrants, attended by campaign group representatives, consular staff, British residents, social scientists, local politicians, and the press. At these meetings campaign group representatives and British con-sular staff spoke formally and the audiences were invited to ask questions and engage in dialogue. After the Elche meeting, Jeremy interviewed sev-eral participants and other campaign leaders. He also traced the activity of some former British councillors via digital archives of the local press. A reviewer of an early article by Jeremy on the anti-Brexit campaign (Mac-Clancy 2019) questioned the nature of his contact with members of these activist groups. So, to clarify, Jeremy interviewed several members of these campaign groups, some more than once. Though he was invited to partici-pate in their gatherings, and their demonstrations, he declined. His relations with them may thus be classed as friendly and highly informative, but he did not get internally involved.

For Fiona, who became an active participant in 2008–09 in several groups run or attended by Britons, her involvement was always clearly defined for the purposes of her research. She ceased participation in those groups following the initial research and, similarly to Jeremy, did not become involved internally with any of the campaign groups.

Our plans of work were approved by our respective University Research Ethics Committees. Part of our ethical practice is to send or show participants drafts of what we write for them to comment on. Jeremy also gave his initial report on attitudes of migrant Leavers and Remainers as a seminar at the Universidad de Alicante in May 2019. All of the interviewees for that piece of research were invited to the meeting, and to a lunch afterwards, to continue the discussion: throughout, Jeremy stressed his keenness for any comments or criticisms they might have. He also plans to give e-copies of this book to all interviewees who would like one. While we, as co-authors, are responsible for everything we write, we regard this practice of sharing drafts and final texts with our interlocutors and seeking their responses as a means to avoid gross misrepresentation, as a gesture towards moral equity between interviewer and interviewee, and as a recognition of fieldworkers' debts. These reasons are particularly relevant in this ethnographic case, since each of us occupies sometimes an interchangeable personal position, between residents and researchers, in our respective fieldsites.

Our structure

In the next chapter we set the ethnographic scene: for both fieldsites we first provide sociohistorical background to our respective areas, then ask, who moved when, why, how, and what have been the consequences? In Chapter 3, we follow the migrants into the town-halls, tracking why they stood, what they encountered, and what, if anything, they achieved. Brexit is broached in Chapter 4, with Jeremy's analysis of the disputes and common ground between Remainers and Leavers, as well as the logical grounds for little hope of dialogue. His work, among migrants in Spain, is then complemented by Fiona's subsequent, parallel querying in France. In the chapter following, we turn to more public politicking: the rise of anti-Brexit campaign groups in both our areas, probing their formation, mode of action, and effectiveness. Since so much of their interaction is enabled by the Web, one particular online meeting place is then examined and assessed for the varied functions it fulfils: informative, affective, socialising. Chapter 6 is a tale of citizenships, rights, and registration: how they can be acquired, lost, fought for, or avoided. We deviate briefly towards the dark side of lifestyle migration: the unregistered, who choose to shy from bureaucracy. We conclude with a glance at latest developments and general comments on approaches to political agency in affective times.

In sum, our book is a comparative study of political subjectivities, of migrants' spirited navigation of politico-civic terrain: their different strategies, aims, and successes as municipal councillors; their affective engagement during the Brexit process; the rise of the anti-Brexit campaign groups, especially their online activity and its partial successes; their renegotiation of their status as citizens and residents at the very moment that Brexit threatens that. In a phrase, our book critically examines the unexplored zones of political agency by British migrants faced with radical civic re-definition.

Notes

1 Earlier versions of particular sections of this book appeared in two of Jeremy's pieces: Barth and Brexit, online, on target, in Eriksen, T.H. and Jakoubek, M. (eds) 2018, *Ethnic groups and boundaries today: a legacy of fifty years,* Abingdon: Routledge; and MacClancy 2019.
2 See, e.g. Valencia's land remains up for grabs, *The Guardian,* 23 April 2006.
3 France was a founding member in 1957; the UK joined in 1973, Spain in 1986.
4 Free movement was introduced in 1951 with formation of the European Coal and Steel Community. It enabled the free movement of labourers to facilitate recruitment of coal and steel workers across its member state borders. With the evolution of the EEC, freedom of movement rights were expanded but still tied largely to labour.
5 As part of a national reorganisation of French regions Midi-Pyrénées was joined with Languedoc-Roussillon in 2016 to form Occitanie.

2 British migrants in Alicante province and South West France

In this chapter we set out our ethnographic scenes. We wish to portray the course of settlement by British migrants in these, our extended fieldsites. We want to demonstrate who moved where, when, why, and to what consequence. In each case we begin with an historical sketch of their migratory patterns, and then enquire into their specific reasons for leaving the UK, and for establishing themselves in their chosen locations. We close with a few comparative points.

Alicante and the Brits: a brief history

Spain is 'the most desirable location for . . . Brits living in the EU': this was the 2018 assessment of the UK's Office of National Statistics (ONS), even after it had taken into account the dramatic decline, from 2013 onwards, in the number of British citizens living in the country (ONS 2018:3). However, reliable figures for Britons based in Spain are hard to come by. Those produced by the National Institute of Statistics, a branch of the Spanish Government, are widely acknowledged as a grave under-estimate given, until very recently, the large number of migrants who avoided contact with local bureaucracies (Table 2.1). It is quite possible that by the early 2000s there were over 700,000, maybe more than a million, British living in Spain (O'Reilly 2017:141).

The province of Spain with the greatest percentage of incomers from the rest of the EU is Alicante (Figure 2.1). Easily the largest group among those migrants is the British. In 2010, 15% of the registered provincial population came from other EU countries: 45% of which were from the UK. And the most popular destination for all Britons do who come to live in Spain is Alicante: 34% choose the province as their new home (Huete et al. 2013:337–8). The volume of migration to Spain has been so great, and in some places so concentrated that 17 municipalities in the country have more registered foreigners than natives, and nine of those are in Alicante, mostly inland: in

Table 2.1 Officially estimated numbers of Spain's foreign population from the UK

	1998	2001	2005	2012	2017
Foreign population: UK	75 600	107 326	277 187	397 992	240 785

Source: Data from Instituto Nacional de Estadística (INE) n.d.a., n.d.b.

2015, in San Fulgencio 75% of the population were incomers, in Rojales 72%, and in Lliber 62%; in all three cases, the majority of these immigrants were British (Ortiz 2015; *La Vanguardia* 2018).

Britons residing on the Alicantine coast are nothing new. In the late nineteenth century, the city, connected to Madrid by railway, served as the busy port for the capital and for exporting produce from the province and beyond. Some British merchants worked in this small but international centre of commerce, while their more leisured or convalescent compatriots were attracted by the 'dry but equable climate' of this 'popular health-resort'.[1] In the words of a British traveller passing through in the early 1870s:

> A small English colony exists at Alicante, with a consul, a chaplain, and a pleasant, hospitable little society . . . They told us that if we stayed long, we should learn to delight in the place, and even to think it beautiful; but to us it appeared so miserably abject and squalid, we could not believe it possible.
>
> (Hare 1873:75, 78)

The opinion of an upper-class aesthete, visiting in the late 1940s, was more tempered, but still damning: 'Alicante could hardly be pleasanter, or more delightful, or less interesting' (Sitwell 1950:107).

Latter-day successors thought very differently. Since the interwar period a trickle of North Europeans and middle-class Spaniards had been vacationing on the coast (Macaulay 1949; Carr 1986). But in the late 1950s, the energetic mayor of a small fishing village in economic decline joined forces with local businessmen. In a famous encounter, he then persuaded Franco to allow tourist development of his municipality. The subsequent rapid construction of high-rise hotels, financed by European tour operators, soon turned Benidorm into a major destination for package holidays. It is now a large town, with the greatest number of high-rise buildings per capita in the world. Since the might of its hoteliers stalled any competitive development anywhere nearby, construction companies switched to building second homes, initially for well-to-do Spaniards, later for North Europeans. They first developed the north of the province, mainly substantial houses on large plots; in the 1980s, they moved on to southern

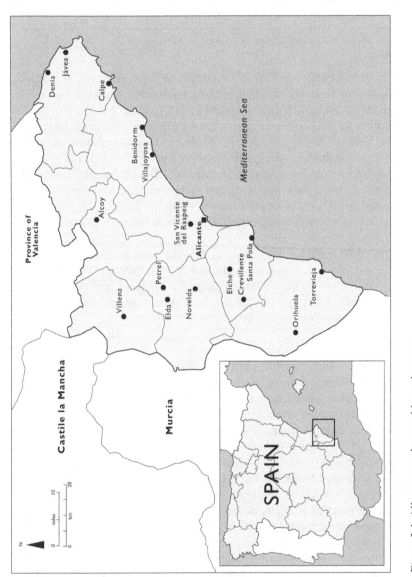

Figure 2.1 Alicante province and its main towns

Alicante (Janoschka 2009). By the mid-1990s, so much of the coastline had been urbanised that constructors shifted to the countryside, building large estates (*urbanizaciones*), mainly of smaller homes on smaller sites.

Some North Europeans who bought second homes went on to live there full-time. Others came directly to live in the area: to the busy coast, the dense estates, or individual homes in the Alicantine interior. Building was so intense that in just the first six years of this century the housing stock of the province rose by 25%, with over 240,000 new homes. During this period the proportion of Britons and other foreigners buying homes in the Alicante area was the highest for any part of the country (Huete et al. 2013:337). Authorities were keen to enable this construction, as the new building and its immigrant occupants promised to stimulate local activity throughout the year, and not just for the high season (O'Reilly 2017:140). Many of the large estates, which may have populations of up to 16,000, were built on the margins of their municipality, some kilometres away from the original village. They can thus act as semi-detached communities, with effectively their own economies and social lives, where British residents do not have to learn Spanish to fulfil their usual needs. O'Reilly, researching Britons on the Costa del Sol in the early 2000s, thought urbanizaciones there comparable to the compounds of colonial days and British 'New Towns' of the postwar decades (Huber and O'Reilly 2004). Many British migrants live year-round in the province; some, called 'summer swallows' or 'yo-yoes', return to the UK in the summer when Spain gets too hot for them. The rise and rise of budget airlines means that though some reside permanently in Spain, others return to the UK occasionally (e.g. annually) while others fly back and forth much more regularly.

The stereotype in the UK of the British in Alicante is of the elderly retired (e.g. Engel 2016): this is a mischaracterisation, as many residents have either full- or part-time jobs, and some are periodic commuters, returning monthly to Britain to work. The age-range of these residents is similarly varied: while nearly 41% are 65 years and older, 54% are aged 15 to 64 and a significant number of those migrated to the area with their young families to raise them there. In contrast with the stereotype, several commentators emphasise the diversity of Britons coming to Alicante: from conventional tourists making a brief vocational trip, to second-homers, to permanent residents; from pensioners to the fully employed; from inhabitants of the coast or the countryside to towns or city-dwellers. They may also question the label of 'lifestyle migration', as that again may mask the diversity of this mobile population and take at face value migrants' downplaying of economic considerations (Huete and Mantecón 2012b; O'Reilly 2020:5).

Lifestyle migrants, by definition, move for the sake of a better lifestyle. They wish to change their mode of living, positively. Though they cannot totally revise their quotidian existence because the ideal of the improved life they wish to lead is part of their home context, they are able to make substantial changes. Many retrain in order to generate an income in their new place of residence, e.g. becoming a yoga instructor, or teacher of modern languages, whether English, German, or Spanish; one, who could not speak Spanish when she arrived, ended up writing a textbook in it for fellow British incomers. Others, however, went further, in effect reinventing themselves. As one couple put it, 'The main reason we're here in Spain: we're anonymous, and can be anybody we want to be'. Several interviewees mentioned some Britons turning themselves into builders, painters, plumbers, decorators, or financial advisers, 'on the plane over', 'all of a sudden', and the costly mistakes they had made hiring these self-defining 'experts'. But if these reinventors had clear financial motives, those of some others tended towards the fantastic. Jeremy was told of an ex-corporal, who today speaks as though he had been a field marshal; a former Royal Air Force auxiliary who claims he was a Wing Commander; and a retired Marine who speaks of his years in the Special Air Service. One interviewee told of a former acquaintance who broadcast that he had owned a very successful but somewhat obscure business in the UK; his teenage son later stated his father had been a warehouse manager for B & Q. Another described a neighbour who claimed to 'have done everything': as a supposed holder of a judo black belt, he said he had trained a James Bond actor and Special Services personnel. We could go on, but the point, we think, is made: by shifting residence to another country, with different climate, language, culture, and system of government, some migrants gave themselves the opportunity to change, sometimes radically, sometimes fantastically.

The process of Alicantine development has transformed both the physical landscape and native society profoundly, perhaps to a deeper and greater extent than in most other parts of Spain. New sources of wealth have created general affluence alongside new forms of social classification. An expanding middle class has dissolved the formerly stark division between landowners and the landless. Some cash-poor fishing families who owned land near the coast have since turned wealthy through skilful management of their holdings. Providing tourist services is now central to the provincial economy, and construction the largest industry in Alicante. Indeed, the leading construction companies grew to be so powerful that they played a major role in directing the economic life of the area. Commentators have said this has fed previously unknown levels of pervasive corruption among Alicantine politicians at all levels, from regional deputies in the national Parliament, *los Cortes*, to village councillors (Aledo et al. 2012).

Brits in Alicante today: why they came, why they stayed

In Spring 2019, Jeremy interviewed 28 British residents, all but one living in the Alicantine interior. They were a diverse lot: singletons, married, and families; three had immigrated with a non-British spouse, two of whom were Asian; one had met their Spanish partner in the country. Only one claimed dual nationality, and that was through his French wife. None was a naturalised Spanish citizen. All his interviewees, bar two, were white.[2]

Why had these Britons chosen to leave the UK, and why had they come to live in Alicante? Most stated they left Britain because they disliked its contemporary development. 'By the time I left, England was not the country I was born in, brought up in, or worked in', said one. Specific reasons included: a retired couple spoke of a growing sense of hostility in urban centres; a homeowner, who had been taunted by teenage neighbours, protested the law sided with burglars; a pair of ex-teachers had perceived a narrowing of educational vision, with too much emphasis on exams; the former owner of a high-street business complained about the increasing length of red-tape. Several wondered if they would now fit in, should they have to go back. Jeremy was told of one couple who did return because of ill-health but came back after two years, as they had disliked contemporary Britain so much. In other words, almost all interviewees expressed dissatisfaction to some degree with the UK; as a place to live, Spain seemed a better bet.

Most said they had come to Spain because the climate was milder and usually warmer; the country was only a short plane-flight away from Britain, enabling quick return if need be, and easy visits by kin and kith. Further, and crucially important for those surviving on their pensions and/or part-time jobs, a better standard of living was possible as the cost of living was lower compared to the UK. Their days were more tranquil, and they had a better social life. As one female entrepreneur, who had felt overburdened by British business bureaucracy, said to Jeremy, 'What's the point of having kids if you can't spend time with them?'

Several underlined their love of Spain as a major reason for immigration. They praised as highly as they could various aspects: Spanish culture, its literature, music, and art, its way of life, the sense of community, religion, its food, its landscape, the light late in the day. Some lauded what they saw as a Spanish sense of family, in comparison to the UK which, according to researchers, suffers from particularly acute ageist stereotypes and a very high degree of age segregation, particularly between grandparents and their grandchildren (Centre for Ageing Better 2020; United for All Ages 2020). Also, Spaniards were thought nicer, friendlier than, for example, French or Italians. One, an interior designer, said, 'I feel a real affinity with the Spanish people, their mañana mentality, with everything about the country'. One artist, bored with 1990s London, was attracted by the slightly anarchistic feel, the rawness he thought Madrid had at the time. Francis Bacon, in a chat at the Colony Club, Soho, had persuaded him that the capital 'was more 24 hours than New York', as he later found.

All interviewees, bar one, lived away from the coast. They chose the province because the Costa Brava (north of Barcelona) and the Costa del Sol (centred around Malaga and Marbella) were too expensive and too noisy. Several at first lived on the Alicantine coast but moved inland after some months or a few years: they found the littoral towns too full of tourists; it was too easy to spend one's time exclusively with other Britons, and very easy to drink too much. There were too many drugs, and burglaries. The beauty of the piedmontese landscape appealed to many; one said he had chosen the Alicantine interior as a World Health Organisation (WHO) study claimed it was the best place for human habitation thanks to its climate, air quality, and altitude. Most lived in detached houses near a town, some on estates, and a few were off-grid in the countryside. All resided within a half-hour drive from an international airport.

The great majority appeared satisfied with the fullness of their lifestyles: socialising, physical activities (walking, walking the dogs, gardening, exercise classes, swimming, cycling, etc.), and organising or participating in communal events, e.g. organised walks and picnics, low-intensity sports (petanque, darts, pool), 'Moors and Christians' fiestas, choral singing, amateur dramatics, a Facebook discussion group, an annual village panto – in English. As one said, 'We get involved in any village activity we can'. Many owned dogs, chosen from the municipal pound; a good number of them volunteer in the shelters or promote canine charities. Cheap but good restaurants meant many can afford to lunch out frequently, which they could not in the UK. A few noted their collective presence maintained or boosted commercial activity in the area: one had set up a beauty salon, another a gym; several worked as teachers of English, and some individuals as masseur, personal trainer, athletics coach, yoga teacher, travel-writer, sculptor, arts therapist. One had made month-long stints in the UK as a live-in carer and knew of at least five women in her town doing the same. Some expressed pride that their children had helped keep local schools open. In one town, several interviewees mentioned to me a particular charity shop, run by a Briton, whose profits funded the panto and kept seven local, Spanish families out of extreme poverty.

In sum, all were well aware they experienced a better standard of life, whether they commuted periodically back to the UK, held down a job in Spain, were semi-retired, or enjoyed the fruits of their pension. Most could afford a bigger house than they could have had in the UK; the others could not have bought a house at all in Britain. Food, medicines, and many utilities were cheaper, municipal taxes lower; there were more hours of sun, and speedier access to health services, of very good quality. Most were socially active and, indirectly or directly, contributed to local services.

No one expressed regret at coming to Spain. Jeremy interviewed a self-confessed quasi-hermit who lived an unhealthy existence in idiosyncratic conditions, and another resident, a widow approaching her eighties, who was bed-ridden at the time: neither wished to go back. Their home was Spain and they did not lament that.

Britons in South West France

The phenomenon of Britons making France their primary country of residence emerged later than in Alicante.[3] While housing purchases are recorded during the 1960s, largely as second homes, it was during the 1980s that more significant numbers of Britons purchased rural, often derelict, properties, renovating them and converting farm buildings into liveable accommodations. With the help of British-run property agencies working closely with the media, rural French properties were marketed to Britons as 'a rural "dream" home' (Hoggart and Buller 1995:179, 1994).

From the 1980s, Britons quickly became the main group of foreign investors in these rural properties, often purchasing realty for which there was little French demand (Barou and Prado 1995; Hoggart and Buller 1995). In the postwar period, the French countryside experienced depopulation, declining local services, and rural poverty. Modernisation and mechanisation of agriculture in the 1950s and 1960s exacerbated this decline as descendants of farming families moved to cities and alternative forms of employment (Bossuet 2006; Clout 2006). The result was an abundance of abandoned and uncared-for rural properties. These left behind old houses in isolated settings or in undeveloped villages appealed to the aesthetic sense of middle-class Britons, particularly those with a preference for home renovation. Similar properties in rural UK locations were largely unavailable or unaffordable. Early researchers labelled the emergent UK to France migration as 'international counterurbanisation' capturing the broadly urban-rural shift of Britons crossing the English Channel (Buller and Hoggart 1994). Quoting estate agents, they remarked that "'the literate found it the fashionable thing to do" as it became the "talk of the cocktail set"' (Hoggart and Buller 1994:181).

Dordogne in the Aquitaine region was the initial hub of British interest. Since the 1960s this area has been marketed to Britons seeking holiday homes (for many, second homes became primary residences). Provence was also popular though considered more expensive, busier. Initially, Britons were reluctant to purchase properties from their fellow nationals but in due course a semi-autonomous housing market developed (Hoggart and Buller 1995). Later British arrivals spread beyond the Dordogne, which one property agent described as 'saturated with British people' (Hoggart and Buller 1994:182). Property prices had also risen, and Britons looked to other regions in search of better economic value. Also, rural areas beyond the Dordogne were becoming more accessible thanks to new motorways and the rise of budget airlines (Teindas 2009).

Today the geography of British property ownership is diverse, but the south west region of Nouvelle-Aquitaine continues to host the greatest proportion of British residents (26%). Neighbouring Occitanie, which stretches east to the Mediterranean, hosts about 17%. Overall, the estimated number of Britons living in France more than doubled between 1982 and 2000, according to ONS (see Table 2.2). In 2017, almost 153,000 Britons were living in France long-term: roughly 68% aged 15 to 64, 19% aged 65 and over (ONS 2018).

Fiona's fieldsite, the former Midi-Pyrénées region, received a significant increase in British arrivals towards the end of the 1990s (Dugot et al. 2008).[4] Between 2001 and 2006, over 36% of the total documented arrivals were Britons (INSEE 2010). The region comprises a range of landscapes, from its mountainous south to the valleys and gorges of the Lot and Aveyron departments in the north. Fiona's focal point was the largely rural department of the Tarn (Figure 2.2). In the words of a geographer who has had a home there since the 1960s,

> The Tarn has so many lovely little rural niches hidden away in its maze of country lanes, valleys and rolling hills. It used to be one of France's "best kept secrets" but it's now been well and truly discovered, not least by the British.
>
> (Wise 2015)

Rumour had it that British media had once referred to the area as 'Kensingtarn', but the link was much disputed by Britons living there.

Many rural villages in the Tarn experienced this rural repopulation by European migrants during the late twentieth century (Bossuet 2006). For example, in one Tarn *commune*, rural exodus led, between 1968 and 1975,

Table 2.2 Estimated numbers of Britons residing in France

	1982	1990	1999	2007	2017
Number of British residents	34 180	50 422	75 546	145 622	152 900

Source: Data for 1982–2014 compiled from INSEE (French institute of national statistics) and for 2017 from ONS

Figure 2.2 The Occitanie region and its departments, South West France

to a decline from 1215 inhabitants to 1017.[5] The nadir was 1990: 910 inhabitants. The trend reversed considerably from 2009 to 2015 with population rising from 950 to 1044: an increase predominantly due to incomers, and particularly the establishment of a British community. Britons, alongside other northern Europeans and French incomers, have helped to improve the quantity and quality of local housing stock, employed local services, attracted visitors and tourists, boosted the number of inhabitants, and created a demand for local services including village primary schools. One Briton told Fiona 'we bought this [chateau] because nobody wanted it. It has been empty for, we think, 50 years'. Another stated, 'our three children helped to keep the school above its closure threshold'. A third revealed, 'I'm told the English saved the local village, which was dead'.

Potential problems were highlighted too. Several implied that Britons had largely influenced price inflation in the local housing market. There were also rumours that some Britons brought in teams of builders from the UK. One French woman told Fiona that she did not mind the British, they were a good thing for the area, but only when they used local services. On the whole, however, there appeared to be little antagonism between British migrants and other inhabitants in this area, echoing earlier findings of Buller and Hoggart (1994).

Unlike many Britons in Alicante who bought newly built apartments or houses on estates, those who moved to South West France tended to choose isolated dwellings or properties in hamlets and villages. One local estate agent told Fiona that 'people wanted a stone house with a pool, within walking distance of a bakery' (he also stated that he did not just sell someone a home, he sold them a lifestyle). Fewer new arrivals seemed to be attracted to the more recently built villas, though several Britons already living in France had downsized and chosen more modern dwellings. Some Britons aspired to build single-storey eco-homes to move into as they grew older but two couples that Fiona knew had struggled to find available local land with building permission. There has been a general tightening of building restrictions around rural villages (less so the larger towns) since the 1990s.

Overall, Britons in France were much less concentrated than in Spain. In Fiona's research area, British migrants never made up more than 18% of the total residents in any one administrative unit. However, many of the villages were home to other North Europeans, notably Dutch residents, but sometimes German and Scandinavians. There were also French incomers (primary and second-home owners) who had lived overseas or elsewhere in France.

How was it that Britons came to be in the Tarn? Fiona's interviewees moved from largely professional jobs and careers, sometimes following retirement, early retirement or redundancy, often looking for a change of life. Most arrived directly from the UK, a few from overseas (former diplomats, those involved with international businesses, one couple who had travelled the world in their boat), and yet a few more from elsewhere in France. The majority underlined the central importance of available and very affordable rural housing: prices were sufficiently low for British buyers to sell their property in the UK, acquire a home of similar or higher quality in France, and still have some money left over. As one couple said, 'What we could have afforded in England, there's absolutely no comparison to what we have here'. Another couple was looking for a change of lifestyle:

> [He] was made redundant in 1999 and we decided that we'd like to have a go at doing B&B and had originally planned to go to Norfolk [UK] but realised that prices had gone up too much for us to be able to afford a decent size house.

Priced out of the UK, the couple used the redundancy money and the sale of their UK home to buy a house in France and set up their business, their surplus cash they invested in a piece of land on which to later build a separate house.

Most Britons interwove the attractions of cheap property with a range of broad factors that included the charms of French rurality and culture, beauty of the landscape; the food and wine; milder weather than Provence, warmer weather than north of the Loire river; cheaper properties than in Aquitaine; less British than in Dordogne; international airports close by; having family or friends in the area; and the perceived need to equalise the demands of work and one's life with a better 'quality of life', 'a better work/life balance', and, for some, an 'escape'. One husband and wife in their forties had left professional careers in London: 'We had for a long time thought that the quality of our lives was unbalanced and had wanted a different quality of life and a different way of living'. For another couple, it was a 'rural escape' they could sustain financially.

For many, a love of France, however mythical or not their image of the country was, underpinned their accounts of why they had moved:

> I always wanted to live in France, simple as that;
> As far as I was concerned, it has been an intention, not even a dream, from a very early age;
> We came to live in a French environment. I want to die in France.

Some emphasised the beauty of the landscape and the delights of French ways: one used French to capture a sense of uniqueness and the exotic: 'c'est un peu caché, un jardin des paradis'. A couple was captivated while holidaying in the country:

> We were seduced by the loveliness of around here, and the really brilliant food and wine. We really got sucked into the whole wine, mushrooms and foie gras, so we just stayed.

Some Britons, younger and older, talked in retrospective, nostalgic terms, seeing the French countryside as 'England was fifty years ago': more 'civilised', with a 'sense of community', and 'family values' which they felt no longer existed to a similar extent in the UK. As one put it,

> Yes, it's like England was. I grew up in a little village in Sussex and it was wonderful. Even though it was wartime you never locked your bike and you cycled into the village. People trusted each other and were polite to each other, and all these things have gradually eroded to nothing now.

Here you still find that civility and the charming French people who are always polite and they always say good day, good morning.

Several mothers also expressed the benefit of this seemingly old-fashioned style of life for their offspring: a 'safer environment' where their children could 'run round and be free, not like in England'. In the words of another:

The main reason we came to France was so the children would have exactly the same upbringing as I had in England.

A further pair of examples illustrates the interlinking of economic and emotional factors; they liked the prices, but they also valued the proximity of kin and kith. One single mother, in her forties, said she bought her house without a mortgage because she valued the 'security of a roof over our heads, come what may'. This would have been 'impossible' in the UK. But she chose her particular area of residence because her parents already lived there. Another mother said she, her husband, and their daughters 'arrived in the October and my parents arrived in the November. My husband's parents are hoping to move out here too'.

An extended case exposes the value of EU citizenship for migrant kin. A father, posted to the USA for several years, went accompanied by his wife, their sons, and her widowed mother. They chose to return to Europe for the sake of a simpler migratory regime, as his wife explained:

Had we remained in the USA [my mother] would have had to keep applying for a visa . . . and they couldn't keep guaranteeing that they're going to let somebody stay on if they're not a direct descendant or a spouse of the person who's working there.

. . . So this was always hanging over our heads and they could have turned around and said, 'No' at any time. [In France] there's no problem. It's the EU and there was no problem with her being here from a legal aspect or having to provide any kind of authentication of what she was doing.

In other words, as an EU citizen with the individual right to free movement, the mother was able to accompany her family without seeking or renewing permission to do so. It gave the family a sense of stability. This reference to the EU was only one of three occasions in Fiona's interviews when Britons referred to the EU to explain how they came to be in France. The second occasion was a Briton who claimed:

I wanted to move out of the UK and live in Europe. I think it was Europe first and then France was a preference.

The third expressed the same: EU first, France second. The limited acknowledgement of the EU in explanations of how Britons came to be in France in Fiona's research echoes the findings of other academics. Drake and Collard (2008:227) found, for example, that 'none of the respondents could relate their own venture to the broader framework of Europe or the concept of a European citizenship'. Among Britons in the Midi-Pyrénées it was not necessarily the case that they could not relate their moves to their EU citizenship rights but rather the rights were taken for granted. This was revealed in phrases such as 'I think of myself as a Brit. who has chosen to live in France and we're all allowed to as Europeans'.

For these rural based Britons, days were filled in various ways. Many worked professionally, six among them regularly commuting to the UK: an airhostess, pilot, desk-based professionals. Four people ran, from France, the same business as they had in the UK. One couple told Fiona 'We realised we can continue to run our businesses here so we thought why do we have to live in a basement flat in Clapham when we can live here?' Others started afresh: international property sales, DIY, construction, massage and wellbeing, pool maintenance.[6] By far the most popular was holiday accommodation as Britons turned outbuildings into rental accommodation. Few people became employees, due to inadequate language skills, lack of necessary qualifications and, most often because it did not fit with the lifestyle they had moved for.

Beyond work, lifestyles included socialising, property renovation, gardening, physical activity, language classes, visits to nearby locations, and organisation of charity and social events. There were groups of Britons who knew one another well, while others were seemingly unknown to their fellow nationals. Some told Fiona that they had never found such an interestingly diverse circle of Britons in one place: they had friends or acquaintances who had travelled the world, were employed in prestigious or unusual places, and were found to be engaging company. A few liked to boast of famous British neighbours, whether they knew them or not. Others very deliberately said, 'we don't mix with other Brits'. A minority had more French friends than British though all expressed knowing at least a few French people. However, there were the occasional examples who regarded their native neighbours as too rural for them. Beyond the two nationalities, a few Britons felt enriched by getting to know co-residents from elsewhere: northern Europeans, Canadians, South Americans. It was striking how most Britons initially defined their friends and acquaintances by nationality. A minority instead defined them first as friends with whom they shared leisure or work activities: through participation in choirs, sports, volunteering, the local council, market stalls, etc.

These British rural inhabitants traversed a range of familial relationships: singletons, male-female and same sex couples, families of one or two parents. They were first and second-generation migrants to France. Some had moved to France with their French spouse; others had met in France their partner, whether French or other national. Three claimed dual (French and British) nationality: one through birth, two were naturalised. In short 'the British' Fiona met were a diverse group in several dimensions, but she was only aware of one non-white among them.

Britons in Toulouse

Fiona, in 2009, also interviewed ten Britons residing in peri-urban Toulouse. Britons living in urban areas of France have been understudied despite reports that they constitute a significant number of Britons in France (Scott 2004, 2006; Puzzo 2007; Buller 2008; Ferbrache 2011a). They explained their moves to Fiona 'for work': all but one had moved for employment reasons: the tenth preferred urban living and was drawn to the vitality of the city. Two of the nine, single at the time and both in their thirties, moved to take positions in the Toulouse offices of their British employers. As one woman explained:

> An opportunity came up with our parent company to manage the International Customer Services team based in Toulouse. . . . I'm based here on a two-year contract.

The remaining eight were employees, and their wives, of the city's aerospace industry. One arrived in 1996, when British Aerospace formed a joint venture with Avions de Transport Régional in Toulouse:

> We moved a lot of people. I suppose we must have moved 200 Brits to work, plus families, so maybe 600 people.

Other Britons had arrived before or after this date, either on permanent contracts or as secondees. Many then stayed in France; the rest returned to the UK or moved to another country. Of those who remained, some had changed jobs; others chose self-employment; a few had retired in France. For these more urban-centred Britons, the main justification for their migration was employment.

All the married employees (all male) arrived with their spouses, and often with young families. They tended to live close to the place of work on the western periphery of Toulouse (Puzzo 2007). There, newly established facilities were tailored to the English-speaking international community: an International School, English pub, grocery store selling British food, English-speaking clubs and societies, a cricket team, and a theatrical group which put on a panto each Christmas.

While the husbands worked, their wives looked after children, socialised together, some volunteered, a few of them worked. Those who became employed taught English, one established a creche, another set up a business to support new international arrivals. Their children were educated in a local English-speaking school; a minority went to French schools, particularly if the move was considered long term for the family. Some of the wives had grouped together to form a social club offering a frequent and regular array of activities for English-speaking women. Some described this group to Fiona as 'a life-line'.[7]

Britons abroad: some generalisations

In both Alicante province and rural South West France, Britons had migrated there to take advantage of cheaper property rates, to enjoy the setting, and to take pleasure in local ways. In all cases the reasons they gave were personal, complex, and multifaceted. These Britons thought the sum of all that would improve their quality of life. In contrast, the examples of Britons living in Toulouse expand the tools required to understand the British population in France. These urban-based migrants (albeit a very small sample) justified their moves for economic reasons that exceed the limits of lifestyle migration, and often the media representations of Britons in southern European destinations. In all cases, whether to rural or urban settings, the majority of people made the move to a specific country, usually one of their choice; only a few migrants to France, and none to Spain mentioned the EU as a reason for shifting.

It is also apparent from our long-term fieldwork, and literature on the topic, that the justifications for Britons moving to Spain and France, economic or lifestyle oriented (or indeed an overlapping of the two) evolve and adapt into reasons for staying, moving house, moving on or returning to the UK (e.g. Benson 2013). What we will demonstrate further on is that the changing political context in which Britons live in Spain and France (as part of the EU) highlights how the migratory outcomes of choices and decisions never remain stable.

The material of this chapter also fits smoothly within the terms of contemporary practice theory. Migrants make use of and develop their various forms of capital when they decide to move abroad. Those who migrated to work in the aerospace industry around Toulouse advance in particular their economic capital. But all, whether urban jobholders, tele-cottagers, leisure-seekers, or pensioners, cultivate their cultural capital to differing degrees: some, for instance, learn a new language or become more competent in one they already speak, albeit in a faltering manner. Migrants' acquaintances with another (French or Spanish) culture normally progresses. In the case of France, this knowledge may be especially valuable as French culture was often seen as sophisticated, and command of its ways as prestigious by many Britons (Thorold 2008; Morris and Compagnon 2010). Some of the migrants to France also develop their social capital, by making valued friendships with co-residents from other countries as well as interesting and lively compatriots. However, the habitus of the migrants we interviewed and of whom we were made aware, in France or Spain, did not change fundamentally, despite differences in degrees of social, cultural, and economic capital, and the extent to which they are mobilised. Their repertoire of habits, behaviours, and skills together constituting their habitus might have broadened somewhat or become more elaborated, but it was not altered radically. The exceptions to this generalisation, the activist campaigners against Brexit, we discuss in later chapters.

Notes

1 *Guía práctica de Alicante y su provincia,* 1908, Madrid: Marzo, A. 34; 'Alicante', *Encyclopaedia Britannica,* 11th edition, New York: Encyclopaedia Britannica, 1:661.
2 We include this detail as Benson and Lewis (2019) highlight that not all British migrants on the Continent are white. In Alicante province, other than the two interviewees already mentioned, Jeremy has only ever seen one other non-white Briton in all his years of visiting the area: both cities and the countryside; businesses, leisure zones, and domestic settings.
3 Britons have a longer history of visiting and seasonally residing in France including, for example, to the Riviera and Pyrenean spa towns.
4 Midi-Pyrénées consisted of eight departments: Ariège, Aveyron, Gers, Haute-Garonne, Haute-Pyrénées, Lot, Tarn, Tarn et Garonne.
5 The *commune* is the smallest of three administrative divisions in France below *départements* and *régions.* Fiona takes the demographic development of this commune as fairly typical for the Tarn. She does not identify it because, in the mid-2000s, she assured her interviewees of anonymity of name and place.
6 British migrant entrepreneurs is the focus of work by French academic Vincent Lagarde (see Lagarde and Di Pietro 2019).
7 We have included this material to underline that some of our research was carried out in an urban setting. Britons in Toulouse were a focus of Fiona's doctoral research (Ferbrache 2011b).

3 Political agency, electioneering, municipalities

In 1992, legal establishment of EU citizenship bestowed on its newly created citizens the right to 'vote and to stand as a candidate at municipal elections in the Member State in which he resides' (Article 22, Treaty on the Functioning of the EU) (Preuss et al. 2003). The Spanish Government implemented this legislation within its borders in 1997, and the French the year after.[1] This was a small but radical move. For the first time migrant members of the EU not born in Spain or France could engage in local decisions that directly affect them and legitimately criticise in a sustained, public manner their local mayor and councillors, and even attempt to replace them in office.

In this chapter we look at the consequences of that shift for local life in coastal Spain and rural South West France. In each case we examine who became councillors, why, and to what effect. We then compare the experience of the two groups: what was common, what was different, and for what reasons. In a closing section we attempt to draw out the more theoretical dimensions of this material.

Alicante province, Spain

The municipal life of Spain is managed by 8,124 town-halls. Their size varies widely: from fewer than a hundred inhabitants to more than half a million. But the majority are small: over 61% of municipalities have less than 1,000 inhabitants apiece.[2] While the number of councillors is officially proportional to the municipal population, in smaller municipalities the ratio is in fact relatively high (Table 3.1).[3] This elevated level of representation can make town-hall concerns and activity a more central part of local life in these smaller electoral districts.[4] Municipal councils enjoy high levels of autonomy and exercise a remarkable diversity of powers: from public security to urban planning, from slaughterhouses to promoting gender equality, from funeral services to co-management of schools and fiestas.[5]

Table 3.1 Examples of municipal representativeness, Alicante province, 2019

	Number of inhabitants	Number of town-councillors
Elche	232 517	27
Castalla	9 880	14
Benejama	1 703	9
Tollos	57	3

Source: Data compiled from the websites of the respective municipalities, all accessed 19 May 2020

From the late 1990s on, Britons, and migrants from other EU countries, have voted and stood for positions in municipal elections. Alicante, as a province with particularly high volume and dense concentrations of incomers, also has among the highest numbers of EU-migrant councillors, and the majority of those foreigner office-holders are British. For example, in 2007 in Alicante three dozen non-Spanish residents, the majority of them British, had seats on town councils (MacClancy 2019:373). Most of these British councillors stood for office in small or relatively small municipalities: only two had anything close to 20,000 inhabitants.

These British councillors are strikingly diverse: in background, political trajectory, and degree of success. Generalisations here cannot be too specific. All but one had at least a full secondary education, a few were graduates. In the UK they had held various positions: company secretary, accountant, restaurant owner, etc., with a disproportionate representation of ex-policemen. Most councillors were at least in their fifties. None had participated in party-based political activity in Britain, though close kin of two had been councillors or mayor back home. All already had some public presence, sometimes in the UK, more often since migration. In Spain, several had won local reputations as energetic activists who boosted charitable organisations, or organised campaigns, for migrant or environmentalist interests; one had set up Citizens Advice Services, weekly in a bar, where he was also treated as an agony aunt.

Their experience of office is very mixed. Several felt marginalised, by fellow members of the local party and within the council, both in municipal matters and linguistically. One felt her successful efforts to raise party affiliation among migrants was then undermined: other councillors for her party feared she would use the boost in numbers to form a migrant faction bent on unseating them. If within the governing group, the incomers were usually given the brief of representing European migrants within the area, but excluded from other business. They considered they were left with specific council business they had expertise in, and which did not interest fellow councillors. Anything else on the municipal agenda was usually kept from them: 'Things always happened without my knowledge or consultation. "Oh sorry! An oversight", they would say. But it happened too often to be only an oversight'. One councillor, who had got his initiatives implemented, said, 'They were initiatives I was allowed to do to keep me from asking too many questions'. Other initiatives of his had been rejected as they 'would have been good for my reputation'; the governing party did not want him too popular. He thought they regarded him as 'a nuisance, and often a thorn in their sides, because I wanted to change things'. In general, these Britons felt they were being exploited for their electoral support from, plus access to and knowledge of a particular municipal population. They were not invited to ponder other matters. They were being used, and contained.

In contrast several considered they had managed to work well with fellow councillors, and had achieved much while in office. They were no longer merely a bridge between the town-hall and the people they assisted; now in the town-hall they could strive to secure reform. Some listed initiatives they had fomented and seen implemented, successfully: applying for multi-million euro EU grants; revamping their town's tourist strategy; running health campaigns; setting up a charitable network to assist the needy; developing town-hall/migrant relations; integrating the socially isolated; winning national awards for mobility strategies; and so on.

At the time of standing, all spoke Spanish at least moderately well; in contrast, some other non-Spanish councillors in Alicante do not gain a command of the language, even while in office. Two of the elected Britons stated native councillors excluded them in meetings by speaking Valencian. Both learnt to understand Valencian. One councillor who works with German and Belgian counterparts said they spoke English to one another. In other words, in some municipalities, council business is now being pushed towards the polyglot (English, Spanish, Spanglish, Valencian, Vanish [Valencian-and-Spanish]), with basic Spanish as lingua franca.

Interviewees stood for a range of parties, with a slight majority on the centre-right. Whatever their allegiance, they talked of national parties in strongly local terms. The municipal chapter was their primary loyalty, its regional branch viewed as occasionally overbearing and self-interested. Interviewees seemed more attracted to municipal personalities than to party stances or national policies. Thus, several had changed parties, more than once, justifying their switches by moral assessment at that time of the representatives involved. For example, one environmental activist started in the Greens, which amalgamated with a left-wing party; she conjointly entered into a coalition with the centre-left. Later she left the coalition and joined a successful motion of no confidence. She then entered a coalition with the centre-right, so winning control of the town-hall. At the following elections she was elected for the centre-right. Her actions are not particularly unusual, whether among foreign resident or local representatives. For each transition, she explained her actions in local, not national terms: some were 'corrupt'; councillors affiliated to one party, now with charges brought against them, had been replaced by a 'new team, people I could trust'. Generally, in Alicantine politics, indigenous membership of parties can be very unstable. *Transfugas* ('defectors, turncoats') crossing the floor is a much-denigrated, much-practised strategy.

Throughout the country, many town-halls have not shown much interest in stimulating migrants' participation in the democratic process. Funding for town-halls is based on the number of people on the municipal register; this is not the case for their electoral rolls. Hence mayors have no economic incentive to increase the number of resident foreign voters; indeed, they may well be unsympathetic to the idea of outsiders influencing local matters, or even not want to facilitate the registration of new voters, whose political leanings might be opposite to those of the group then ruling the town-hall (Rodríguez 2018:8). In many municipalities, the sum result is relatively few incomers voting, let alone standing in elections (Méndez-Lago 2010; Tomé da Mata 2015; Bermúdez and Escrivá 2016). Since most political parties, as a general rule, have not striven to diversify their membership but put obstacles in the way of non-Spaniards entering, or, rising within their ranks, some resolute migrants have stood as independents or formed their own parties (Simó-Noguera et al. 2005:20–1; Burchianti and Zapata-Barrero 2017).

One interviewee said he had approached local branches of national parties only to be rejected: they did not want those they regarded as outsiders. He and fellow migrants formed their own political party, though administrators made their official recognition very difficult throughout this bureaucratic process. In 2010 they finally established Partido Independiente por las Nacionalidades (PIPN). At its height, its representatives were elected to two of the 13 councillorships in its local town-hall, while its busy Facebook page demonstrates its sustained raft of activities and campaigning. Certain comments on the page make clear the desire by some in the party for secession of the large, migrant-dominated estate: electorally, the estate dwarves the original village yet receives a disproportionly low fraction of the municipal budget to fund services. They wish the estate to constitute its own municipality, receiving its residents' taxed income rather than losing it to the village town-hall. This municipal rebordering, primarily for fiscal purposes, is a long-established practice in Spain. Migrants' deployment of this strategy is a further example of their adaptation to their new place of residence.

Most interviewees were surprised by the deeply politicised conduct of municipal business, with accusations, along political party lines, of corruption or favouritism embittering town-hall debates. Some were taken aback how easily partisan interests smothered communal concerns. But since none had direct party-political experience in the UK, their surprise lacked comparative base. Some complained of vote-rigging: the registration of some migrant residents 'dropped off the register'; supporters of the mayor's party were, interviewees stated, allowed to vote though unregistered. Interviewees also described the ademocratic style of some mayors, which might edge towards the dictatorial. One said the mayor came to his house and banged the table: 'This is how things are! This is how things will be!' He was later found guilty on multiple charges of corruption and banned from holding office for several years.

'Corruption' was a much-repeated concern of interviewees. Since its perceived scale is an evolving composite of political, judicial, economic, and media factors, Jeremy cannot judge its incidence, only report its perceived prevalence. Alicantine sociologists say over the last 30 years the devolution of powers from higher to municipal levels of government has enabled corruption to spread (Huete and Mantecón 2012a:91). One interviewee stated that in the 2011 elections, the till-then governing party and the opposition forces won four seats apiece, making him 'the key'. Shortly after, one telephone-caller offered him a car; another said, 'Go with us, and by the end of your term, you'll be a millionaire'. Some British councillors have adapted so well to certain local practices that they, in turn, have been accused, sometimes formally, of corruption. One of the accused resigned so that any of his planned, further municipal actions could not be questioned, and to reinforce his point that the mayor, already facing multiple charges, should himself have stepped down long before.

Nepotism was a further concern. Two interviewees from one municipality said all eight councillors constituting its town-hall governing body are close kin or in-laws, with municipal contracts going to their relations, in-laws, and friends, in a stereotypical clientelist manner: large contracts are waved rapidly through meetings; smaller contracts can lead to surprisingly large bills. Throughout Alicante province, the continuing succession of court cases against municipal corruption, widely reported in the local press, suggest these practices are widespread. According to one British ex-councillor,

> I have witnessed so much corruption here that some politicians treat it as a hobby. Some councillors treat the Council workforce as their own and get them to do their gardens, paint their houses, etc. The workers say nothing for fear of losing their jobs. Politics in Spain is dirty. It needs to be cleaned out.

The Alicantine sociologists refer to a 'deficit in the quality of local democracy' (Huete and Mantecón 2012a:89).

Anthropologists of Europe speak of small-town inhabitants forming 'moral communities' (e.g. Heiberg 1989; Sorge 2009). We may be observing much the same here. As British consular staff underlined, 'There is a large "grey area" between what is illegal and what is *enchufismo* ("plugging-in"), i.e. helping your family and friends get jobs and make money'. Corruption appears accepted, so long as it is kept within bounds. British migrants repeatedly told Jeremy most locals, though friendly, still regard them as 'fair game'. However, his field data suggest migrants are not singled out: any outsider from beyond the locale may be taken advantage of. Spanish colleagues of Jeremy did not deny it, just downplayed its incidence, i.e. these moral communities usually exclude others, whether from other lands or provinces; and when assessing municipal actions, locals live in a constant tension between state-defined legal codes and what they will accept as tolerable practice.

In Jeremy's interviews, complaints about corruption, nepotism, and local forms of democracy segued with moral assessments grounded on 'fairness' and 'justice'. The secretary of one residents association said, 'I'm not going to stand for office. I just want things for the people here'. The councillor who set up Citizens Advice Services said he had done so 'Because I'm a mean Scotsman' who thought local service providers overly greedy, 'just taking money from the expats. This is unfair. Lots of things are unfair in this world, and I'd like to rebalance it'. The PIPN Facebook page speaks of making 'creative and meaningful steps towards political empowerment of the expat community'. Guided by their 'moral and ethical concerns, our intellectual contributions and our strength of numbers', they 'want to help to give our community the voice it needs and deserves'.[6] Its founding President told Jeremy its ethos was 'fairness, equality, openness for *all* the residents of the municipality' (original emphasis).

Rural France

French municipal elections take place at the level of the commune. There are approximately 35,000 communes, varying in size by number of inhabitants.[7] Seventy-five percent of communes have less than 1,000 inhabitants, though they range from fewer than 100 to more than 300,000. Commune size determines how many councillors make up the municipal team (see Ferbrache 2019b). Given that the proportion of councillors to number of inhabitants tends to be quite high, particularly in smaller communes, local government in France is highly accessible to franchised individuals, even more so than in Spain (Collard 2013). Each council is headed by a *maire* (mayor), holding considerable power to manage the local unit in accordance with national policy. However, development of intercommunal structures (e.g. syndicates, community of communes) has reduced some of this power by encouraging communes to work together around issues such as refuse collection, water management, and road repair.

Britons participated in local elections in 2001, 2008, and 2014, and were most prevalent in councils of smaller communes, i.e. those below 3,500 inhabitants (Collard 2013; Ferbrache 2019b), reflecting the residential pref-erence for more rural properties. Those Fiona interviewed were councillors in populations ranging from 180 to 4,200 inhabitants. In small-sized communes such as these, participation tends to be apolitical, though this does not pre-vent the possibility that teams of candidates will share political perspectives (Collard 2010). The number of Britons standing and elected as councillors has risen considerably across three terms, and in 2014, 896 Britons became council members (see Table 3.2). While this number appears relatively small considering that there are roughly half a million municipal councillors in France (and also in comparison with estimated numbers of Britons residing in France), Britons constitute a significant proportion of non-French council-lors, particularly in smaller communes (Collard 2013). Moreover, there has also been very little official promotion of electoral opportunities for non-French EU citizens, just as this has been lacking in Spain.

Fiona interviewed 13 individuals who stood for election, most of whom gained office, and two who were invited to stand but declined. She carried out the interviews in 2016, mostly with solo participants. On three occasions, partners sat in and contributed to the interview, while in another instance a councillor's wife intermittently joined the conversation while cooking in the kitchen where the interview was taking place. In one interview a former and current councillor were present simultaneously, and another where the councillor took Fiona to the *mairie* (town-hall) to meet the maire. Everyone was interviewed just once.

The majority of interviewees were in their fifties or sixties; two respon-dents were in their forties when elected; one in their seventies. At the time

Table 3.2 British municipal councillors elected 2001–14

	2001		*2008*		*2014*	
	<3500	>3500	<3500	>3500	<1000[1]	>1000
Number of commune inhabitants						
Number of elected Britons (*standing*)	*	16 (*79*)	41 (*108*)	405 (***)	772 (*1133*)	124 (*396*)

Key:
[1]1,000 became a new threshold ahead of the 2014 elections
*Data unavailable

Source: Data compiled from Collard (2010) and Ministère de l'intérieur (2018)

of the interviews four were fully employed (farmer, consultant, financier, self-employed), and the others described themselves as working part-time (an English teacher, consultant) or retired (semi-, early- or entirely). Their former UK professions were diverse across teaching, nursing, telecommunications, financial, managerial, and governmental work. Three Britons had children under the age of 18. The children had previously attended their local village primary schools before progressing to secondary schools in nearby towns. Most interviewees were the only Britons on their council, but three councils had two Britons, and another had three. No individual had prior municipal experience in the UK. However, one had been engaged in social activities of his village before moving to France. No one had been involved in what they considered to be activist organisations in either country at the time of interview.

All electoral candidates had been invited or encouraged to stand by the incumbent or prospective maire. This is characteristic of the personalised nature of elections in smaller French communes (Collard 2013). They said they were chosen because they spoke French, at least to a good level, and were known to engage with locals. One stated that she was seen regularly around the village walking her dog and that during those walks she would often stop for short chats with other residents. Another explained that he had become friendly with the maire through attending all the commune's events. Almost all interviewees were already active in other commune groups and activities: two served on their school Parent and Pupil associations, one was President; another played and coached local sport; one served on the festival committee organising events and hosting social activities; someone else had established a local choir; another was part of the patchwork group; one managed two holiday rental properties belonging to the commune; two ran their own businesses and engaged frequently with local organisations.

Some said the maire was expressly in favour of including British people. One, for example, was invited as, 'There were more and more Brits coming to our particular commune and [the maire] thought, quite rightly, proportional representation . . . that we ought to have a Brit. on the local council'. Similarly, in other communes:

> Our maire was quite keen that there was a representative from the English community because at that time we were about 15% of the total population;
> [I was recruited to] represent the interests of and to communicate with the English speakers.

In practice few councillors would find acting as a conduit between the council and British residents became a particular part of their mandate.

In most cases, being a councillor involved taking responsibility for one or two public services in the local area. Portfolios held by individual councillors included council finances; schools and playgrounds; culture and heritage; sustainability; and the community magazine. One claimed no particular responsibilities, since everyone contributed to everything in his commune. Others became involved with their commune's water and electricity services, tourism, sustainability strategies, local planning, road repairs, street lighting, and sport and recreation facilities. To manage these activities and contribute to broader commune business, each person gave their time and effort, which ranged from one meeting per month to more frequent arrangements when specific projects were carried out. Councillors with responsibilities such as roads, water, electricity, and sustainability might also attend intercommunal meetings. In one instance, a participant explained,

> I did go on training courses to learn how to be on these committees, which was very useful. So I spent a day with the electricity people and a day with the water people and that gave me a lot of confidence in what I was doing.

Others laughed off the idea of training: 'no, it really wasn't that well organised'. Overall, respondents were active in overseeing the public services and general wellbeing of their commune and the people within it. In the words of one, 'It really is the minutiae of daily life', a banal politics.

Banal, perhaps, but not without conflict. Though councillors did not divide along political party lines, Fiona was told on several occasions of major differences and factionalism within the council and commune. Three councillors defined factions in their village 'between people who have supported the last maire and the ones who support this maire'. One woman found herself excluded for standing against her maire: 'the maire and the maire's wife and a couple of other people with whom I'd been friendly, started ignoring me. It was all very much "you stood against us so we won't have anything to do with you"'. Another explained:

> one of our members turned out to be a sort of hidden supporter of the last maire, undercover. She caused so many problems, until it got too uncomfortable for her and she resigned.

A third Briton explained that two opposing electoral teams were in conflict over claims of electoral fraud. A regional electoral commission was called on to help resolve the matter, but mistrust persisted: 'the maire had his finger in a lot of pies and contacts with politicians further up the political ladder, and strings had been pulled'.

Conflict also arose due to 'village politics', as one Briton called it. He referred to differences and disputes 'between French people, historic families of the village'. One councillor told Fiona how 'inhabitants took particular sides because that's what their parents and grandparents had done'. Another suggested that 'the disagreements go back and back, and nobody really remembers what it was all about'. However, these Britons were able to avoid such 'historical' conflicts by positioning themselves as 'newcomers' and 'outsiders'. In contrast, one suggested that her position as a newcomer (and non-French councillor) gave her the ability to intervene:

> I can say things on a council meeting, "Why don't you shut down that polling station, it seems a lot of waste of manpower". And then everybody looks askance and says "You can't say that". And everybody agrees but they don't dare say it, because it would be too political because they've all been living there years and years. It doesn't matter to me, and for them to say "well she doesn't understand". It actually brings the subject to the table without anybody saying, "Oh well, it was him or it was her that said that". That would get around the village and then they get black balled by the whole of that village. It doesn't matter for me.

To participate fully, Britons told Fiona 'you have to have the confidence to do it, and the language', the latter because all council business is conducted in French. All British councillors spoke fluent or good levels of French. In practice, four of them found the speed, accent, and jargon of spoken French challenging. One elaborated: 'particularly if they're excited and heated in conversation and then I find them almost impossible to understand'. Most who were not already fluent speakers valued the challenge and the benefits. In addition, two said that lack of fluency had prevented them from making significant contributions.

Socio-cultural belonging

The geographer Lynn Staeheli has written of notions of democratic community as 'rooted in some form of commonality'. Whether one stresses what is 'common' or the striving for 'unity', she argues a sense of community is constructed via shared experience, common values and a common concern for place (Staeheli 2008:8). All three factors were evident when Fiona asked interviewees about the meanings they gave to their municipal participation. One was motivated by the opportunities 'to share in the day to day running of the community'; another wanted 'to integrate' with the people; a third valued 'being able to participate and being able to be part of the local commune'. A further interviewee said she participated in 'the hope of trying to forge some deeper relationships . . . and to show we're willing to take part'. One respondent said he enjoyed,

> just getting to know a pretty large proportion of the population, to have a real sense of community. When I go to the market on a Saturday morning, there will always be a number of people that I know, and who will come up for a chat.

For another, his time on the council was about finding common values with fellow residents:

> It was really important getting to know the people on the council better and understanding what their concerns were. They were not my age (we were the oldies of the group) so we didn't necessarily have the same interests.

In other words, the council helped to bridge perceived differences through working together.

Becoming a councillor was also a conscious means to develop a deeper attachment to the place, by engaging in more diverse and, for them, more significant networks of exchange:

> I enjoy the fact I can be actively involved in this small community. If I live somewhere, I want to feel that I'm part of that structure, part of that place.

Several spoke in similar terms: 'to contribute to the life and development of the village in which I live'; 'I like getting involved, it's interesting, and you give something back'; 'it's the ability to take part really . . . to do some good and make a difference to some of the things in the village'. Only one articulated a desire to combine a sense of shared experience with common concern for the future: 'sharing in the day-to-day running of the community and involvement in plans and investment projects for the future'.

When Fiona asked Britons about key contributions they had made, several referred to visual changes in the village and to longer-term alterations which clearly demonstrated improvements to the local environment and its services. One walked Fiona through the centre of the village, proudly pointing out the trees she had selected for planting. Another laughed as he explained 'my legacy':

> Inside toilets.
> I have been banging on about this all the time on the council and then when they finally agreed they were going to extend the hall, they agreed to put toilets and handwashing facilities inside.

A further respondent underlined his involvement in a tree-planting cere-mony, a local custom where each new member of the council plants a tree at the start of their mandate. Since the tree would outlast his term on the council, its continued existence and gradual growth would display his con-tribution and presence in the village for a long time hence. This was not so much an individual achievement, more his personal participation in a com-mon, and communal mode of memory marking. Other personal contribu-tions and achievements ranged broadly: 'a sense of economy'; 'a charging point for an electric vehicle'; 'refuse sorting'; a walk for locals and tourists; recruitment to staff a village café.

Fiona's interviewees revealed that these various changes to the commune contributed to the sense of place which they felt, and valued. As one coun-cillor said, 'to improve the quality of life in the village and its economy'. To their pleasure, they had actively, visibly, perhaps even memorably par-ticipated in the ongoing maintenance and amelioration of their village and to the wider benefit of the community there. They were making a place for themselves, in their place of residence, which others could see, use, and appreciate.

However, not all participants could easily identify specific contributions that they had made. One councillor explained:

> I wasn't doing anything. I did ask about it and they said 'don't worry, when they need you they'll ask you', and as yet I have not been asked to attend anything.

The limited contribution had left her feeling 'a bit embarrassed'.

A second woman was very modest about her achievements. She showed Fiona copies of the council magazine that she put together twice yearly and then said 'I'm just happy to let the ones who are in a better position to get on with it, those who have a better understanding of the system'. There were also those who did not feel that they had made any achievements at all: 'None, to be honest. None'. Another suggested that 'successful initia-tives are limited because the maire resists change', while a further example confessed, 'I am generally disappointed with my achievements compared to what my expectations were and I have found it incredibly difficult at meetings to make any significant effective contribution'. Reasons for lim-ited contribution included a council resistant to change, lack of confidence to share ideas, lack of French fluency, and a sense of banality towards the matters being discussed:

> I wouldn't go as far to say that I had any feeling of achievement. I did my bit. It was a frustration. I very quickly learned, like a lot of things in

French life, it is heavily laden with procedure and bureaucracy and so on, and you just have to get on with it.

Three of five councillors stood for a second or third term. Two moved to a new house in a different commune which prevented them restanding; one was adamant he would not repeat the experience in his new location: 'I think once is enough. It was mundane to the extent that it wasn't an experience I would like to repeat'. One councillor starting in 2001 hoped to step down at the end of her mandate by recommending someone in her place. She was advised otherwise: 'the maire said to me "look, it's taken everybody seven years to get used to you, so don't rock the boat now"'. Of the councillors elected for the first time in 2014, several aspired to stand in 2020, some felt it was 'too early to know'. This turned out to be a wise comment in the wake of the June Brexit referendum, which took place a week or so after the interview. Post-referendum, one woman told Fiona 'not being able to be on the council, it's like having a slap in the face'.

Comparing experience in Spanish and French municipal councils

The experiences of British councillors in these neighbouring countries appear strikingly different. But perhaps these contrasts mask what may be similar. Let us see.

In Spain, the councillors speak of personal scruples, when defending their various initiatives and occasional campaigns. These elected representatives attempt to justify their actions and aims in terms of what they might class an everyday ethics. They propound a morality whose appeal and persuasive power is based on an implicit claim: their ethics is so commonsensical, so universally acceptable that no further grounds of justification are necessary; hence their unquestioned concern to denounce corruption, nepotism, and other municipal abuse. They utilise an 'unofficial' ideology, one not subject to the same stringent criteria of internal coherence as the carefully thought-through social theory of professional party politicians. For councillors, unexamined notions of fairness and justice are good-enough yardsticks; popularly accepted modes of evaluating ideas and behaviours whose deployment should prove relatively uncontroversial. In effect, they argue what kind of municipal candidate could have any chance of electoral success if they dared to openly oppose the upholding of fairness and justice?

Their words have bite because, in their view, self-interest, factionalism, and political partisanship threatened to dominate town-hall business. Their counterparts in France told a very different story. Instead of heated debates about the award of municipal contracts, they talked of a cordial ambience where consensus was usually achieved without too much effort. For British councillors in France, their work in their council consisted mainly of serving and managing the commune, for the sake of its general improvement. They did not suggest there was significant resistance to change; rather they spoke of working together relatively harmoniously to manage the daily functioning of the commune in an efficient, effective manner. The only fault-lines within their meetings appear to be ones internal to the commune and not grounded on economics or avarice. When Fiona asked about the politics of the local council, one interviewee replied, 'Do you mean big 'P' politics or little 'p' politics? Because the only politics relates to historical local disputes between certain families'. Another revealed, 'It is not political, it is personalities really'.

At first glance, this Hispano-French difference appears to be one of municipal size. Jeremy's interviewees came mostly from municipalities several-thousands strong, with some having populations close to 20,000; most of Fiona's interlocutors lived in communes with less than 1,000 inhabitants. In fact, on further examination, this does not appear a relevant difference, as in Spain the town councils of even very small municipalities, those with only a few hundred inhabitants, are usually divided down party lines. A key dimension here is thus one of political culture: in Spain, party affiliation is a crucial identifier of elected representatives even at the lowest levels of government; in France, party membership is of little significance in smaller communes where local matters take precedence.

British councillors in Spain and France also had somewhat different ultimate aims. Those in Spain wanted to bring about important change: to counter injustice and correct poor management. If they also wished to integrate more closely into local society, they did not openly refer to it as a personal priority. The social benefits of councillorship were only mentioned in some interviews, and were presented as a by-product of one's term of office, albeit a pleasurable one. Whether this was foreseen or intended was not stated. Fiona's work in France demonstrated almost the opposite. Britons there stood for election primarily for social reasons: to expand their acquaintance and get involved in local projects (Ferbrache and Yarwood 2015; Ferbrache 2019b). The social orientation of their municipal activity was also suggested by the political ignorance of some councillors. Fiona found evidence of a few who, initially, were 'not really clear on the process'; Drake and Collard (2008) found much the same in northern France.

In neither area did British councillors speak of their work in terms of citizenship, rights, their nation-state, or the EU. In fact, a few days after the referendum one woman discussing her role on the local council made it very clear that she had no idea it was contingent on her EU citizenship. Their focus, and their activity, was bounded by the local; and their moralities, though universal in potential application, enmeshed with the local. The regional, the national, and the supranational were not mentioned in our interviews. In France, British councillors wished to enhance their social interactions and attachment to place. At the same time, they wanted 'to contribute', 'to give back': they recognised their role within a network of exchanges and wished to develop that. Moreover, it was a way to perform their vision of, and further participate in, a just society. In Spain, their counterparts upheld similar ideals, but here some put up front the need to actively fight injustice. And to achieve that they intermeshed more strongly in local society. In other words, councillors in both areas upheld comparable dreams of how people should live in communities, with the key difference that those in France put sociability first with local participation as a means to that end, compared to some of those in Spain who gave priority to securing justice while acknowledging the concomitant benefits of sociability.

Whatever the specific priorities of either group, both examples highlight the local polity as a central site in which belonging can be negotiated and maybe deepened. In Staeheli's terms, these non-Spanish and non-French councillors manifest that though village life is to an important extent conditioned by higher levels of management, an engaged, active sense of local belonging 'is part of daily life, something we enact, even as it is part of a broader system by which order is maintained . . . an order that enables us to go about our lives' (Staeheli et al. 2012:631).

Practice theory and town-hall habitus

Bourdieu's approach, born out of his study of Kabyle rural society in 1950s Algeria (Bourdieu 1977), and modern versions of practice theory fit cases where change is gradual or modulated. They dovetail with an analysis of maintained, but developing, municipal practice, where foreign residents exploit their social capital to win seats, and develop their cultural and local capital, e.g. by learning to understand Valencian, French, and local ways.

In Spanish town-halls, the routines are humdrum, though the practices may be lively. Councillors' habitus has long accommodated fiery debate, petty tyrants, secession, nepotism, clientelism, switching parties, and other forms of ill-regarded but much-pursued municipal behaviour.[8] The election of British residents has not altered significantly this habitus; indeed some have been accused of corruption themselves. British councillors may propose and see implemented innovative initiatives, but this is a long-established part of municipal government. Native councillors may work to restrict their effectiveness, to limit their activities, and to fragment their power blocs but, again, this is in the nature of time-honoured interfactional competition: i.e. the elected foreigners have both learnt to adapt, and to adapt to, municipal ways. The habitus evolves; it is not questioned fundamentally. Similarly, in France where routines and practices are banal and mundane, habitus evolves, but slowly.

Janoschka, who fieldworked in Spain during the mid-2000s expropriation campaign, asked whether the 'temporarily radicalized habitus' would turn into a permanently reconstituted one (Janoschka 2011:229). None of the British Jeremy spent time with mentioned this campaign; when asked, most confessed ignorance or very mild, distanced knowledge of it. This suggests that foreign residents' sense of activist history is shallow: no incomer councillors have yet retained their posts for a third term, while many migrants return home on widowhood or impending infirmity. Jeremy's field data also suggests the mid-2000s radicalisation of habitus was indeed temporary; it did not extend beyond the successful achievement of the campaign's aims. The organisation formed by the protestors achieved its aims and ended. Its lessons have seemingly been forgotten.

Until EU laws were introduced into national legislation, it was commonly accepted that non-national residents were not to criticise Spanish politics publicly. In France, non-nationals had long been excluded from any level of franchise. But once the respective governments had accepted that non-national EU-member state residents could vote and stand in municipal elections, Britons and other migrants became able to exercise hitherto-unknown rights: to become political participants in a European country in which they had not been born or raised, but in which they resided. The bond between nationality and the ability to voice local political opinion had been broken.

In Spain, for the first time, non-native residents could stand on soap-boxes or rise in town-hall meetings, and publicly damn the conduct of local Spanish representatives without being accused of meddling in other people's business: from 1999, it was their business. In France, this change proved particularly controversial even at a level where participation was almost apolitical (Arrighi 2014). While France enacted a significant step by enfranchising EU-citizens, their involvement was at the same time curtailed by additional legislation. To prevent their influencing national sovereignty, they were prohibited from holding office as maire, or deputy maire, and from participating in any process connected with the election of senators (Ferbrache 2019b).

This enabling shift in the link between nationality and political participation is the key, perhaps revolutionary, change here: in terms of practice theory, it was more a change of field than a radicalisation of habitus. In other words, the verbal strategy or disposition was not new. Habitus did not change. But those who could practise it had altered. The field had been expanded to include migrant residents. Legitimate political activity within an EU country was no longer dependent on the actor being a citizen of that country. Though confined to a municipal stage, migrants could now be political agents.

Notes

1 The history of implementation in Spain is explained by Rodríguez (2013) and in France by Arrighi (2014).
2 https://es.statista.com/estadisticas/633516/numero-de-municipios-segun-numero-de-habitantes-espana/ (Accessed 19 May 2020).
3 www.lamoncloa.gob.es/espana/organizacionestado/Paginas/index.aspx (Accessed 19 May 2020).
4 This comment is based on Jeremy's observations of town-hall activity in Navarre, the Basque Country, and Alicante province, in periods between 1985 and 2020.
5 «Ley 7/1985, de 2 de abril, Reguladora de las Bases del Régimen Local. Título 2. Capítulo 3.» (Accessed 19 May 2020).
6 Quotes from PIPN Facebook: www.facebook.com/PIPN-Partido-Independiente-Por-Las-Nacionalidades-189052697809847/ (accessed 30 May 17).
7 It is difficult to determine the precise number of communes, as different sources provide varying figures.
8 Evidence for this statement comes from Jeremy's experience of rural fieldwork in Spain, which started in Navarre in 1984 and continued into the late 1990s; and from his discussions about this chapter with two Spanish anthropologists based in Alicante province.

4 Brexit, a referendum, a declaration of values

Few felt indifferent about the referendum. In our field-areas the result upset many, pleased some. Throughout, it excited strong reactions, powerful passions. Brexit divided; it united. It made many migrants reflect on who they were, who they wanted to be, and what they desired of the world. In this affective process it shattered some socialities, reaffirmed others, and created new ones. It also strengthened some political subjectivities, sharpening their profiles and deepening their bearers' allegiance to them. We must not exaggerate. Some relationships endured, and some valued that duration: they did not let Brexit corral conviviality.

In this chapter we investigate the sentimental economies created by the Brexit debate, and interrogate migrants' attitudes towards the binary positions created by politicians. For the migrants we interviewed, we look at: the fundamental values upheld; the opinions expressed; whether they coalesced in a coherent manner; and how those who positioned themselves at either pole learnt to live among one another. We look in particular at Britons in Alicantine Spain, then compare their responses with those from migrants in France. The material of this chapter thus acts as a foregrounding for our examination, in the following chapter, of the pro-Remain migrants' organised campaigning.

Brexit: opposed logics?

Over spring 2019, in Alicante province, Jeremy interviewed 28 migrants from the British Isles. Three lived on the coast, the rest inland. Everyone had a clear opinion about Brexit. Of the 26 Britons interviewed, 20 were for Remain, six for Leave. Almost all interviewees stressed they held to their position without any qualifications whatsoever. Four of the 26, though claiming to cleave to one side or the other, held mixed positions which could be seen as somewhere in between the two poles.

Remainers

Jeremy groups the reasons they gave for voting Remain into: (a) familial, (b) international trade, (c) belief in larger units, (d) belief in the EU as a beneficial organisation to be a member of, and (e) identity.

Familial: several said they thought of their adult children in the UK and their future prospects, especially offspring who ran their own businesses. They judged that Brexit would endanger their profitability, and maybe their survival.

Trade: several worried about the effect on commerce. One argued that so much trade today was international, but one could not think that the pattern of British trade after leaving the EU would be similar to that previous to entering it. Workers would lose out. A former accountant was concerned about the size of the UK as a commercial unit: 'If the EU was a company, England would be just a subsidiary, with nothing to gain by leaving; only by staying and trying to change it'.

Belief in larger units: worries about the comparative size of the UK went beyond the economic. Several argued that the larger the group, the better. There was greater chance of successful defence of group interests, and representativeness. One couple stated it was good having many people behind the creation of regulation: 'The wider you go the more representative you can be'. Several linked belief in bigger units with conviviality, underpinning that link by the general aim of moving towards a greater, more harmonious whole. 'We ought to work towards a world where we all work together. With more tolerance and greater understanding'.

Belief in the EU as a beneficial organisation to be a member of: if the success of larger units is grounded on the goals of conviviality and harmony, then Britain needs the EU more than the Union needed the UK.

For several, conviviality and harmony in broader units, multicultural ones, dovetailed with mobility for all across the Union: being able to decide where to live in Europe was 'a more exciting possibility' than having to remain in one country. All interviewees were, by definition, migrants, and some saw the flipside to British emigration – immigration into the UK – as an economic benefit of being in the EU: 'the jobs the English don't want to do. How is the NHS going to survive?' Harmonious conviviality and general mobility, whether of emigration or immigration, fed off one another: thanks to the new freedom for EU members to wander around its territory, 'people were gelling'. Enabling easy movement between countries both demonstrated and served as a reminder that, in the words of one, 'This world belongs to us all'. Only one

Remainer spoke in negative terms about immigrants to the UK. Just as Collins and O'Reilly found in a pan-European survey of Brits abroad, many Remainers among Jeremy's interviewees regarded freedom of movement not merely as an opportunity for individuals to exploit but as a valued identity expressing 'a sense of who and how they are' (Collins and O'Reilly 2018).

Any large organisation has its downsides. Four interviewees acknowledged the EU did not work perfectly. Two of them admitted they supported Remain, purely for reasons of self-interest: for one, the EU was 'making so much out of England'; another worried about the power of lobbies. However, none of them saw imperfection as a reason for exit but rather for internal reform. Moreover, remaining in the EU was a powerful guarantee for peace, maintained since World War Two, the longest period in centuries without armed conflict in Western Europe. In sum, as one put it, it was 'much more healthy to be part of Europe'.

> *Identity:* all interviewees identified as British. But several expressed that was neither their exclusive, nor predominant mode of identification. They were not British-and-nothing-but- British. One had 'always felt more European than British, which had past its days of glory'. Another had always seen herself as European rather than British. She, with her parents, had toured much of the Continent as a child. For some Remainers, the referendum result and the subsequent debate shattered core beliefs about their home country, exposing them as illusions. One, eliding 'England' and 'the UK', stated, 'I thought I knew what England was about. Now I don't'. In the words of another: 'Please stop it. It's breaking my heart'.

Leavers

The reasons Leavers gave for voting can be grouped in a strikingly similar fashion, as (a) trade, (b) belief in smaller units, (c) belief in the EU as a damaging organisation to be a member of, and (d) identity.

> *Trade:* two said the British economy would not crash. They acknowledged 'bumps' in the first EU-free years but affirmed the country's economy would emerge as strong as ever. The war years had demonstrated the English were 'very good at austerity'. The City of London would continue to be a European centre of finance because it was too well-established. After all, England had been trading for 1,000years before joining the EU. One admitted membership of the EU did benefit trade, 'But they'll be there anyway'.

Belief in smaller units: one said he always believed in running one's own show, whether it be household or country. Charity began at home, he stated, before damning the ring-fencing of British international aid by David Cameron when Prime Minister. This inward-facing dovetailed with another's view that England had done 'too much for people coming into the country', over-generosity leading to an overburdened NHS. One Leaver used the village he lived in to argue for this upholding of lesser groups:

> Pueblo del Moro wants to be Pueblo del Moro. Its people want to retain autonomy. They don't need anyone else. Without identity they're awash, they're nothing. The EU brings a conformity they don't embrace, attitudes they don't follow.

On this logic, expanding the Union, another contended, towards the East was mistaken: 'Their customs and traditions didn't mix with the EU. Though the EU was trying to make it happen, they couldn't'.

Belief in the EU as a damaging organisation to be a member of: for Leavers the EU had become a corrupt operation of unelected but powerful bureaucrats who wasted money yet never got their accounts signed off. Freedom of movement was a fine principle, but not if it let refugees in or allowed 'terrorists and Islamic fundamentalists' to travel without hindrance. In the pithy phrase of one, the Union was by now 'the longest exercise in collective stupidity'.

Identity: a former soldier turned successful businessman said he, his father, grandfather, and other family members had all fought for their country. To him, sovereignty was sacrosanct, his number one. But it had now gone. Millions had fought and died for British sovereignty; however, as two Leavers stressed, Britons were not federalists.

Brexit: the result, and the resulting cross-fire

For some Remainers, the very idea of a referendum was a mistake: 'The Government should sort it out amongst themselves'. Many expressed their reaction to its result as radical 'disbelief', 'stunned', 'absolutely dumbfounded', 'shocked', 'bloody silly', 'mesmerised'. Several stated they felt sick, cried, or burst into tears. One Leaver perceived the day differently: 'Almost everyone shocked. But many of them said, "Good!"'

It is easy to exaggerate. It makes for a neater, more dramatic picture. Not all were so emotionally struck by the vote. For one townsdweller, 'There was a lot of shock, but not much reaction. The pueblo was tranquilo'.

Several said they had been surprised by the way some friends had voted. They quickly learnt to keep their opinions to themselves, for at least a year after the referendum. But discussions were rekindled once scare stories started to appear in the press: what would happen to their pensions, their access to healthcare; would they need to obtain a residence permit, a Spanish driving licence?

We should expect those on either side of the Brexit divide to exaggerate or lampoon the attitudes of the other side. This is common in meetings between people with very different visions of the world (Boon 1983). Here, the relevant question is in what terms they characterise those in the opposite corner. In interviews, Remainers crabbed, in particular, Leavers' view of history, their idea of Great Britain, and the sources of their views, while Leavers criticised Remainers, above all, as ill-informed.

Some Remainers criticised some Leavers as people who 'still think we have loads of gunboats'. They saw Leavers' view of Brexit as a 'retrogressive step, going back to a time no longer existing'. Moreover, as one underlined, Leavers' historical memory was very selective: 'Leavers have forgotten the state of England at the time of entry to the EU': the three-day week, devaluation only a few years before, the UK was 'the poor man of Europe'. One summed up Leavers' attitudes as 'Britain's best, no matter what happens'. Several Remainers pinned their hopes on the younger generation, for whom the Second World War was unlived history.

Several were unhappy by Leavers' lack of consideration for national diversity, both ethnic and geographic. In their opinion, Brexit had given not just the far right, but a broader swathe of people permission to air 'xenophobic nastiness'. Some confessed to being 'very upset by the racial hatred Brexit has excited'. Two referred to the effect of Brexit on sub-units of the country. Both spoke of Northern Ireland: 'What really galls me is the total disregard for the . . . border. . . . A hard border again: it doesn't bear thinking about'. One pointed out that Leavers made no mention of Scots' views.

Many Remainers saw themselves as informed, and Leavers as ignorant. One complained about a couple of in-laws who had voted Leave because of two television programmes they'd watched: 'They'd not thought further than their nose. "Rule Britain" shit', which at root she suspected racist. Another said Leavers talked about regaining sovereignty, but did not provide the evidence. Informed Leavers were crabbed as not thinking the issues through: to some Remainers, Leavers had not considered what life would be like for their grandchildren; also, they failed to value easy movement within Europe.

Some interviewees made the more general point that people, of whatever stripe, were more ignorant at the time of the referendum. They blamed the government for lack of voter education. For Remainers, the underlying implication of the economic forecasting studies produced since the referendum is that Leavers, not Remainers, would change their minds. Jeremy found no evidence of this.

Some Leavers were similarly critical of 'Remoaners': the only reason some residents had voted Remain was that they were frightened. According to one Leaver, worry about losing pensions and healthcare cover was 'arrogant nonsense' fanned by 'fear-mongers': the Spanish Government did not want to lose a significant sub-population which maintained so many local jobs. 'Brexit bad for the economy? That was said by people who didn't understand economics'.

Several Remainers aimed their ire at government, Leave campaign leaders, and the newspapers. They blamed a generation of governing parties, of whatever colour, for using the EU as a scapegoat. Whenever a political decision was unliked, they said, politicians blamed it on EU regulations, if they could. In Spain when a construction or road-building project was part-funded by the EU, that fact was displayed boldly on large billboards: in the UK, the same sign would make no reference to the Union. And, for these Remainers, this sustained sidelining or damning of the EU prevented pro-Remain politicians from praising the Union for the ways it had benefited Britain. Some stressed the manipulative, lying style of Leave politicians: the result was 'the knee-jerk reaction of voters pushed by people with their own agendas, thinking of their own pockets'. Many also blamed the press, especially tabloids aligned with the right.

In this atmosphere of mutual recrimination, how did those couples whose partners voted in opposite ways cope? 'We don't talk about it', two couples said. One spouse stressed it was important to place the topic within broader contexts, to downplay the matter: '*Jeremy!* There's wine to drink! Coffee to be had! Life to be lived!'

If, in social encounters, partisans of both sides choose to stay mum on the topic unless with known co-believers, it is on the Web that the vitriol can become plainer. In the mildly regulated environment of the Internet, where interaction is still speedy, but not face to face, some residents choose to state bluntly what they feel about supporters of the opposite party. Several residents said they had, by mistake, joined a discussion group supporting the opposed side to them. The number of abusive posts they soon received was so great they quickly left. The coordinator for the Facebook group for one Alicantine estate said the volume of posts about Brexit became so large, and the tone of many so charged, that she felt obliged to shift the topic into a separate sub-group. In this e-forum, which Jeremy joined, people appeared to speak more freely than his interviewees had done. Some Leavers openly criticised the number of immigrants – 'changing the genetic make up of the UK for ever' (in a UKIP Warwick poster put up by a Leaver) – and their supposed ability to claim benefits. They also questioned the democratic nature of a second referendum: 'You can't move the goalposts just cos you lost. Suck it buttercup'.

In response, Remainers stated Leavers did not distinguish between asylum seekers and migrants, whether legal or illegal. One lamented that he had 'found out that a lot of people in the UK are racist, which is a shame'. Some stressed the age profile, and thus claimed lack of representativeness, of Leavers, 'foaming through their clenched dentures'. They also queried repeatedly what kind of United Kingdom Leavers wanted, given their lack of concern for the future of Northern Ireland and Scotland, as though they were true 'Little Englanders', as that would be about all they would be left with: 'And they have the cheek to call Remainers "traitors"'.

Leavers and Remainers were at least on the same Facebook page. They did hold some things in common: upholders of either side claimed those of the other used 'lies' to make their case. Some branded their opponents as 'ignorant', 'thick', 'stupid', substituting 'derisory drivel' for factual evidence. Both accused the other of unreality: 'beyond reason'; guilty of a 'mass delusion', acting like Flat Earthers, they 'voted for a unicorn'. It is easy to overstress this division. Rogaly, who researched a small British city, found that many inhabitants actively resisted the common racism by practising a category-crossing, non-elite cosmopolitanism (Rogaly 2020). We can see hints of this here, as there was the occasional, but repeated, desire to retain balance and a convivial co-residence on the estate where they all lived.

Brexit: the future

No interviewee, whether Leaver or Remainer, said that Brexit had changed their minds about going back to Britain. None wished to leave Spain. Some were adamant: 'This is my home'. 'Nothing would make me go back to England. I feel comfortable and at home here. There are no instances under which I'd go back'.

Many stated Brexit had only reinforced their determination to stay put. Some Remainers argued against return because, 'Brexit has so polarised the population', shrinking people's compassion and increasing levels of racism and bigotry. A few said that, if forced to leave, instead of returning to the UK, they would move to another warm, equally cheap country: Thailand and the Philippines, for instance. They had realised the benefits of cheap living in sunny climes and were not going to surrender those easily. They would rather become serial transnationals than reluctant returnees.

The only reasons, people stated, which would oblige them to return to the UK were all financial: the freezing of their pension; loss of free access to healthcare services; a significant fall in the exchange rate. At the meeting we held in 2018 in Elche, the second city of Alicante province, the majority of participants said that even if becoming Spanish citizens gave them permanent residence and access to services, they still did not want to surrender their British citizenship. In 2019, however, many interviewees stated they were ready to forfeit their original citizenship for the sake of staying in the country. 'We'll do whatever it takes'. Only one interviewee bluntly affirmed they would never consider it.

Brexit: what chance of a dialogue?

On the whole, Remainers had an expansive, forward-facing vision of the world, one whose populations were or should be members of the largest groups possible. Members of these groups would strive to embody a pervasive conviviality and harmony. The EU (the group including the British) would be structured by a creative multiculturalism within an ambience of productive tolerance. Remainers recognised they were British, but this was neither their sole nor most significant mode of identification.

In contrast, Leavers saw themselves as upholders of representative democracy framed by a relatively small, once homogenous unit with a strong economy. Its continued existence was partly justified by those who had died for it. To a significant extent, their vision was inwardly focussed and retrospective, of a stable society whose members could recognise one another as fellow nationals in a relatively unproblematic manner, and who had all contributed equally to the commonwealth, according to their means. For Leavers, being British, or more accurately, their understanding of being British, was by far the most important group identity they upheld, beyond that of their individual families.

Some of the examined online material, discussed in the next chapter, did suggest there were racist dimensions to this discourse. Its extent is difficult to gauge. Almost all Jeremy's interviewees were White, some of whom hinted that the topic was too touchy even to raise. As one stated, 'You've to be really careful when talking about migrants, as people'll call you a racist'. Only two Remainers did talk openly about the matter. One complained strongly of the Government giving preference to incomers over Brits: 'foreigners' in the UK 'stealing *my* pension, because they're claiming child benefit'. When Jeremy responded that government finance was not structured that way, she asked him to explain. The other did not give Jeremy permission to print their comment about immigrants to the UK. A third interviewee, a Leaver, was more circumspect. They prefaced their comment with 'Not being racist', then stated 'The UK has done too much for people coming into the country'. They praised Spain, where an immigrant needed to have a NIE (*Numero de Identificación Extranjero*), issued by the National Police, 'in order to do anything', in contrast to Britain which did not appear to have a similar system.

At least one British commentator has framed Leavers in the UK as an odd mix of revolutionaries and restorationists: free-market trailblazers, and those who wish to see their dream realised, of return to a postwar Britain with a stable social order (O'Toole 2019). In contrast, Leavers in Spain spoke predominantly in socio-political, rather than economic, terms. The 'bumps' on the path were a necessary price borne, much less to advance an extreme neoliberalism, than to achieve their idea of British society. Since the referendum, a common, though criticised attempt to account for the result was that people in 'left behind' places, marginalised, post-industrial parts of the UK, had voted Leave for the sake of an alternative, almost out of despair (Edwards n.d.) This was clearly not the case among the British of Alicante: none of the interviewed Leavers were strapped financially, and the few economically pressed interviewees, bar one, argued strongly and coherently for Remain.

If we compare the two groups, most members of each uphold a certain vision of the world, as it is or should be. Both visions are undergirt by abstract conceptions and principles, but by different ones: expansiveness vs. centripetality, multiculturalism vs. cultural coherence, forward-looking vs. historically grounded, unrestricted mobility vs. relative stability, globally oriented vs. Anglocentricity. There were four exceptions to this binary division: the interviewee who complained about the cost of supporting migrants to the UK affecting her pension; another pensioner who was opposed to the EU but, as a British migrant in Spain worried about their financial future, voted Remain out of 'self interest'. A third person admitted she had been inclined towards Remain, for the sake of her four adult children, all of whose jobs might be affected by Brexit. The fourth espoused consistent Leaver views throughout their interview yet called himself a Remainer. When asked why, he replied, 'In my heart I'm a Leaver. But my brain says Remain'.

Members of both sides were united in the centrality of the Brexit process, though they evaluated it in opposed manners. They agreed implicitly that this debate was one worth arguing, that it was an opportunity to stake out one's principles. Leavers and Remainers were also equally disenchanted with their condemnation of the parliamentary proceedings over Brexit. They were united in their judgement of MPs as irresponsible, more concerned with infighting and scoring points than resolving the key issues.

Within both camps, some grounded their views on the fact Britain was an island. The geographic metaphor was common, but those on each side worked it in their own way. Some Remainers saw this positively, e.g. 'We are always English. We have always had sovereignty. We are an island, therefore difficult to take over, so it can't be more over-run by migrants'. Others saw it negatively: 'Britain is a stupid . . . island out in the Channel: but it needs Europe to be something, desperately'. One Leaver argued in terms of size: Britain was too 'small' to accommodate further incomers, and therefore should not be so 'generous' to migrants. One Remainer summed up the difference between sides in islandly terms: Leavers, 'island mentality'; Remainers, 'gypsy pirates mentality'; i.e. nationally fixed vs. transnationals; homebound vs. voyagers; those who stayed on their ground vs. those open to new winds.

At the same time, geography was a potential cause for confusion on both sides. Almost all Remainers and Leavers referred in our discussions to 'England', as though it were equivalent to 'Great Britain', even in the midst of this fractious debate about national destination. For them, the regions of Scotland, Wales, and Northern Ireland were not important fractions of the national whole, whose geographically framed opinions on Brexit might well be significantly different. Even a Scots Leaver I spoke with did not bother to make this distinction, as though the issue overrode important nationalist aspirations. One interviewee, in her late sixties, argued that this elision of the two terms was a consequence of their upbringing: it was only in the last decade, with the ascendancy of Scots nationalism within its homeland, that the distinction had become important. There was similar confusion with 'Europe'. As much a political as a geographical term, Europe is the only continent which is not a separable landmass. The United Kingdom can leave the European Union by vote; it can only leave Europe by geographical re-definition.

In the conversational midst of these opposed visions, compounded by cartographic confusion, several did recognise the difficulty of comprehending their opponents. One Remainer couple felt the argument on their side was so right, they 'couldn't imagine people wouldn't see the benefits of being in the EU'. 'I can't believe the stupidity of Leavers. (The referendum) was really about immigration but know it's dodgy to say that'. Some, mindful of manipulation during the campaign, branded Leavers 'brainwashed'. But as one insightful Leaver observed, 'When you ask Remainers, "Why did you vote that way?", they can't really tell you, and it's the same for Leavers'. We take her incisive comment to mean that the arguments from either side were not engaging, but sliding past one another. For several, whether advocating Leave or Remain, commitment to their position was so firm that it could blind them to understanding others or that, failing to recognise the fundamental differences between the two camps, they traded deprecation over the effects rather than the root causes of the debate. For though both desire a utopia and view that of the other side a dystopia, Leavers and Remainers start from similar bases but interpret them differently and structure their arguments according to opposed values. Thus, usually, they are not talking with each other but at one another.

Leave/Remain is a manufactured difference, the product of public disagreements between professional politicians. Several interviewees said they had been quite happy with the ways things were, until they were forced to think where they stood on the issue. None of the interviewees had held public office, nor belonged to a political party beforehand. One declared he wasn't 'political' before the debate had started, and had never voted consistently for the same party. In other words, Brexit had politicised the previously apolitical.

Politicisation on the issue, however mild or strong, did not compute into collective action. The great majority of interviewees were ignorant of the anti-Brexit campaign groups; the rest, bar one, were knowledgeable of but apathetic towards them. Only one of the 20 Remainers said he had joined some of these groups, and he had only learnt of them via a spontaneous Facebook search, not from any contact, digital or physical, with members of any campaign group. Several said individuals had come to give public talks in a local bar about what Brexit means, but these meetings had had no further consequence.

In sum, in Spain, the Brexit debate created two temporary groupings, of opposed views of the world, strongly upheld by people who are politicised but not collectively organised. These world-views, however coherent or not, are ethical stances, providing their upholders with an identity they are comfortable with. Both Leavers and Remainers are passionately committed to their particular position, and often derogatory about those on the opposed side. These are positions so profoundly felt that Leavers, for instance, are prepared to put their conception of nationhood over economic concerns; they place their political beliefs and national visions before their own personal futures. They support Leave though, they fully recognise, it may well impact on their continued residency in their new home of choice. This can make it difficult for Remainers to understand them. In contrast, Remainers are ready to laud multiculturalism, despite some sociological evidence questioning its claimed benefits (Putnam 2007; Abascal and Baldassarri 2015). Remainers uphold a novel system of government, which though still evolving, continues to suffer from a democratic deficit and burdens taxpayers with a relatively inefficient, and therefore costly, bureaucracy. In the words of one,

> I agree (the EU) has failings, but despite the strength of powerful lobbies, it has put through some extremely important legislation with regard to the environment and in general is a force for good.

Of course, Remainers might ask opponents for an example of a supranational administration whose efficacy was deemed near-perfect.

Some Remainers see Leavers' 'wartime lessons' as indicative of how backward-looking they are, how ignorant of contemporary global markets. At the same time, some Leaver veterans see their experiences of war as validating their views, as precious examples younger generations have not lived through, and are all the more impoverished for that lack. Given their mutually contested approaches to history, we can state Leavers find hope in the past, Remainers in the future. What chance the present?

Responses from rural France

Britons in Fiona's field-area were also agitated by the result, and a good many of them very agitated. In late 2019/early 2020, she tailored discussions with migrants she had already worked with in 2016 and 2017. She also shared with them, for their comments, a draft of Jeremy's work on attitudes to Brexit held by their counterparts in Spain.

The Britons in France agreed that attitudes there were 'broadly similar' to those in Spain. In the first months after the referendum, the online reactions of Remainers were parallel in emotional tone to those of their counterparts in Alicante: 'This decision has completely knocked me sideways'; 'I wish the OUTERS knew the pain and anguish they have caused'; 'I am so angry, and I'm a placid easygoing sort of person'; 'I spend endless nights awake with a feeling of uncertainty and worry'; '[The possible consequences are] the stuff of real nightmares, or certainly insomnia. What if, what if'.[1]

Generally, migrants in France who supported Remain said they did so for familial, political and personal reasons, similar to those given by Britons in Spain: 'Leaving would create a very unclear future'. In 2016, 'uncertainty' was the word most frequently expressed by French-based Britons who responded to Fiona's survey: 'It will make the future uncertain to have no status'; 'There will be an extended period of uncertainty for us'; 'Main problem at the moment is uncertainty regarding the consequences'. Others described the impact that Brexit had on them as 'cataclysmic', 'very unnerving', 'bringing instability', 'insecurity' and 'a state of personal limbo'. These Britons lamented the loss of EU citizenship ('I'll no longer be a European citizen') and its associated rights. A reoccurring concern was their future status and, as one said, 'my right to live in France'. Several asked rhetorically 'What if I have to leave?' Like those Britons in Alicante alienated by change in their country of birth, they were concerned about returning to a UK they 'no longer have anything in common with'.

The few Leavers known to Fiona or her contacts appeared to be dispersed among a population of Remainers. They reasoned in ways similar to those from Alicante: the UK should run its own economy, security and immigration. One dismissed concerns about the future, claiming to be 'fully cognisant of the possible consequences, and am proud and happy to have [voted Brexit]'. In 2016, the Leavers Fiona encountered had expressed little concern about the future: 'They're not going to throw us out, are they?'; 'If I have to go, I have to go'. In contrast, one Leaver stated adamantly that he would not return to the UK and was quietly confident of spending the rest of his life in either Spain or France. In 2019, some, it emerged, had voted Leave for negative reasons. One Remainer couple knew of people who had voted Leave 'as a protest' against UK Government policies, then 'immediately switched to Remain' when the result was declared. One Leaver told Fiona that he moved to France to 'escape the UK' and had 'no empathy for Britain anymore'. Voting Leave would sever those ties more permanently. His unexpected reasoning, which neither of us encountered from any other interviewee, was not to liberate the UK from the shackles of the EU but to deepen the sense of distance between the Continent and his country of birth. He voted Leave for anti-British reasons, not for patriotic ones. This contrarian stance reveals how diverse the reasons for voting Leave could be.

Remainers' reactions were emotional, at times hostile towards 'the other': 'A close friend voted to leave. When we meet up, I simply have to block this from my mind. Otherwise I despise him'. Remainers who thought anyone already residing in France would vote to remain, perceived Leavers as uncommitted to France: 'If you voted Leave, why are you living here in the first place?' Some, nurturing generalisations which both justified their position and fed their bile, thought Remainers tended to be longstanding and permanent residents, while many Leavers were but part-timers, splitting the year between the two countries. As in Alicante, Britons in France tended to avoid discussion of Brexit, unless sure of their interlocutors' opinions. One couple said that at the start of one supper, the host had declared, 'There are Brexiteers and non-Brexiteers here tonight. The topic is banned from conversation'. One interviewee remembered that at another dinner the subject was similarly taboo: 'No politics, no religion, no sex, no Brexit'.

Structuring sentiment

Whether in Alicante province or South West France, British migrants emoted, often strongly, about Brexit, their passion girding their political subjectivities. If we compare these divided Britons in both countries, the structure of their opposed arguments leant strongly towards the congruent. On each side of the argumentative divide, stances were framed ultimately in common ethical values and grounded on fundamental notions of history, collective identities, and destiny. In this strongly affective economy, most expressed a blinding commitment to their position, whichever it was. They disciplined themselves not to discuss the topic with others, unless known co-believers. And their attitudes, usually deeply held, tended not to shift, but to become yet more entrenched over time: as one pair said to Fiona, 'We haven't really seen people switching sides'.

Migrants' affective positions could orient more sharply their moral compass and restructure their social life. But, in both countries, it took the organising abilities of certain Remainers, as energetic as they were outraged, to channel the passion to broad political effect.

Note

1 These quotes are taken from the FFU Facebook page discussed in the next chapter.

5 Getting agitated, together, about Brexit

From their very beginning, anti-Brexit campaigners have come together in their upset and anger. They have shared their indignation, and turned it, digitally, to a productive end, ultimately to a continental boundary. The affective dimension to this politicising process played a patently central role, which must not be downplayed. And an incisive way of studying the political channelling of this sentimental economy, of going beyond anecdote, is studiously reading through the online discussion sections of anti-Brexit and pro-rights organisations. In this chapter, we first examine the rise, organisation, and effectiveness of these groups; we then look at the affective consequences of their digital activity, particularly through the prism of one specific French-based campaign group. Municipal councillors and campaigners are then compared to highlight what is distinctive about the activists. We conclude with Isinic reflection on the creativity of their activities.

Responding to Brexit

'I feel as though I have been shat on': the blunt statement of one intervener from the floor at the Perigueux meeting. Others felt as strongly, but were more scathing than scatological. For many Britons in Spain and France, their first reaction to the referendum result was numbness. They say that soon morphed into an extended, emotional process they explicitly compare to grieving. This 'bereavement', as they term it, could last three weeks or considerably longer for many. It was a state mixed with worry, fear for the future, and 'severe depression'. One campaign group representative spoke of its members' 'shock, disbelief, anger' at the result.

In some parts of Spain, these sentiments were heightened by the sardonic reaction of migrants' native friends. One said the day after the result a village acquaintance approached him in the street with a condescending smile, 'But Peter, you are still here! When are you going home?' Another, a town-councillor, said that on the same day the local mayor, a friend, knocked on his door to say, with a smirk, and in English, 'Bye, bye, Jim. Bye, bye'. Though sardonicism was not universal, no one Jeremy spoke with or knew reported sympathetic interest from their Spanish neighbours. In France, one woman told Fiona, 'Oddly, I have experienced some anti-British sentiment for the first time in over 20 years', but this was a unique example. In Fiona's field-area, Britons were consoled by their native, and other European, neighbours. One said, when she had felt 'very low about Brexit', that a French friend had been 'extremely encouraging and optimistic'. The writer Adam Thorpe, based in a Cévennes village since the 1990s, states that the day the result was announced, most locals were sympathetic and keen to commiserate; only a few took jibes at the unhappy Remainers in their midst (Thorpe 2018:194–9).

In both countries, deeply upset Britons shared their anxiety, grave concerns and anger with similarly afflicted friends and compatriot acquaintance, both face to face and digitally. Some of the more energetic agreed to organise.

Learning to campaign

From 24 June 2016, the campaigning groups began to emerge, first between co-residential kith, next by word of mouth, then by email and the creation of online-mediated groups. It is important to emphasise from the beginning how much campaigners' activities have been enabled, indeed boosted by social media, especially Facebook and Twitter. For this struggle is the most successful UK-oriented, transnational popular political protest so far against the actions of the governing party in Britain, and it is so effective and well-organised because, partially but crucially, much of its activity takes place online. Most of these groups have at least one, if not several IT-savvy members within their inner circle; the websites of many are sophisticated and kept rigorously up-to-date; their Facebook pages are equally active fora for disseminating information and exchanging opinions. Given the feelings generated by the referendum, these fora may fulfil further functions, as discussed later.

Many groups grew remarkably quickly, primarily thanks to the Internet, e.g. the week after the referendum some like-minded British migrants in Almería, southeast Spain, met and agreed to canvas support for an organised group, which they dubbed Europats. Within six weeks, its organisers had developed a database of about 1,000 email addresses; by June 2017, 6,000. In France, Remain in France Together (RIFT) was launched the day after the referendum. Within two weeks they had 1,000 members and in June 2016 their numbers had also swelled to 6,000. By 2020, RIFT had more than 17,000 members. Here, IT is making the velocity of interaction a social factor of its own, its temporal affordances enabling choice of political trajectory.

Some groups are nationally framed, e.g. Bremain in Spain, RIFT. As online-oriented communities, however, campaign groups are not necessarily confined to national boundaries. The constitution of some is explicitly transnational, aiming to represent British migrants in more than one country e.g. Brexpats-Hear Our Voice. Some, mindful that the EU also wishes to negotiate the position of EU citizens in the UK, seek to act for both Britons abroad and EU migrants in Britain, e.g. Expat Citizen Rights in Europe (ECREU), Eurocitizens, and Fair Deal for Expats. In early 2017, 11 groups formed a broad EU-wide coalition: British In Europe (BIE). It claimed approximately 30,000 members by June that year.[1] Most put the particular priorities of their individual groups temporarily aside for the sake of exploiting the power of agglomeration. Members of the coalition's Steering Committee began to teleconference at least once a week to decide strategy and discuss and agree on actions; members were also in daily contact via communications software. Information is disseminated to member groups via Steering Committee members, and to the wider membership and the public via its website and via social media. The coalition is successful at representing the interests of the majority of its members; so much so, that at the Perigueux meeting the representative of one organisation publicly questioned his group's continued, independent existence.

Leaders of some of the Spanish-based groups accept the majority of their membership is regionally confined; this can lead to recruitment flatlining. Leaders of geographically wider groups see the resulting organisational diversity as a richness, not a weakness: the more regionally oriented groups can maintain activity at the local level; the wider groups can invest more energy in broader, transnational initiatives. A more local, Hispanocentric group, such as Europats, tends to focus on working with their local town-halls, municipal commonwealths (*mancomunidades*), and deputies to the national Cortes, seeking and often receiving support from these different levels for their claims, as well as contacting British Members of Parliament (MPs).

The pre-Brexit political experience of the activists living in Spain is mixed. Some had none previously; one leader of a major group stated that current affairs had not interested her at all until the referendum. However, unlike the municipal councillors of Chapter 3, several leading anti-Brexiters do have British-based experience of activist campaigning. Many of them had careers in public service: the NHS, the police, municipal administration; some have received honours for their distinguished public service. But remarkably few of them had participated beforehand in campaigns in

Spain. Thus, for British citizens, standing for the town-council and setting up major Brexit campaign groups appear to be relatively separate trajectories, so far. This difference is not absolute: a sole British town-councillor did attend the Elche meeting: she had recently got her town-hall to declare concern for its British residents, and had then personally handed the declaration to the Prime Minister of Spain.

Some of the leading French activists have former and ongoing campaigning experience in France: one concerning welfare, another the loss of Winter Fuel Payments for Britons abroad, a third Votes for Life (a campaign to secure life-long franchise in the UK for Britons living abroad).[2] Campaign leaders have some tailored experiences: one worked for the Conservative Party for 30 years, someone else volunteered in British community organisations in France. Others had worked in law, management, media, personal relations. Former and current roles are highlighted on many of the campaign websites, conveying a sense of the groups' collective skills and lending a degree of legitimacy to their organisations. There were others, however, with no prior experience, but they 'oppose Brexit' or are driven by wanting a 'Post-Brexit fair deal'.

Whatever their organisers' level of experience, their concerted actions appear to achieve results, keeping their concerns high up on the Brexit agenda of the British Government. Their struggles have been performative and include: intervention in legal actions in London's High Court (October 2016, led by Gina Miller) and Supreme Court; block-emailing the Department for Exiting the European Union (DexEU); issuing an Alternative White Paper on the EU (Notification of Withdrawal) Bill; lobbying MPs, lords, and MEPs; maintaining contacts with embassies, consulates, academics (such as ourselves), and the media; co-launching a European Citizens' Initiative to guarantee EU passports for post-Brexit Britons; co-organising, participating in, and speaking at anti-Brexit demonstrations in London, Madrid, and other cities; January 2017, being accepted as representatives of British migrants and transmitting their concerns to the House of Commons Select Committee for Exiting the EU.

BIE, whose strategy team is mainly comprised of lawyers, has broadened its activities to a wider set of audiences at the local, regional, national, and European levels. Their doings include meetings at the European Parliament in Brussels; in Spain with a range of political parties in the Parliament and Senate; and in France with the French National Assembly, the French Ministry of the Interior, as well as other national governmental departments and British embassies and ambassadors in these and other (mainly western European) countries. In particular, they have held meetings with the Chair of the Brexit Select Committee in the House of Commons, and with the European Commission's chief negotiator of Brexit, Michel Barnier.

All too often politicians engage in games of 'smoke and mirrors', where what is said is frequently more strategic than directly informative. Statements can be made for a plethora of reasons, and interpreted in an equally diverse manner. These tactics can make it very difficult to gauge with any certitude the effect of these groups' campaigns on the Brexit negotiations. Also, a certain scepticism is necessary when assessing the self-made claims of groups about their effectiveness. That said, and as far as we can judge, the sustained actions of the activists, via social media, in face-to-face meetings, in the streets, do appear to have had a noticeable steer on the course of these debate and negotiations. This seems especially the case in the first years after the referendum, when it appears the anti-Brexit campaigning by Britons abroad caught the British Government unaware. Examples adduced by the groups point strongly that their words have had weight: for instance, evidence supplied by Fair Deal for Expats was mentioned by High Court judges in October 2016, concerning the triggering of Article 50; information supplied by Bremain was re-stated in the House of Commons, and one of their lobbyists quoted in debate in the chamber. In March 2017 the coalition was pleased the report for the Committee on Exiting the EU accepted several of the migrants' concerns. Further, as one Bremain representative emailed Jeremy in 2018, 'We have considerably influenced the content of the EU proposals on citizens' rights – it is evident every time we submit new information, as the proposals have changed before our eyes, sometimes even using the exact same language'.[3]

Though many group representatives have previous campaigning experience, they were surprised at the cynicism, duplicity, and exploitation they were experiencing from governing politicians. In early 2017, David Jones, Minister of State at DeExEU, told the House of Commons, 'We have engaged a range of stakeholders, including migrant groups, to ensure we understand the priorities of UK nationals living in EU countries'. Since none recalled speaking to Jones, campaigners responded by block-emailing DeExEU with their concerns. One activist said MPs had told representatives of her group, to their surprise, 'You've moved away. Why should we bother?' Also, the governing Conservative party had told them that a Votes for Life Bill, guaranteeing the electoral rights of British migrants, would go through, but then failed to ensure that it did. In April 2017, in conference-call meetings in the British Embassy, Madrid, and the British Consulate, Alicante, Jones and other spokespersons of Her Majesty's Government (HMG) emphasised they wished the campaign groups to make repeated representations to the Spanish Government to support their demands. At meeting's end, the spokespersons thanked group representatives 'for telling us things and ideas we'd not have thought of'. One campaign leader saw this as DeExEU relying on the groups for information, and wanting them to do its work for Ministry ends:

softening up the Spanish Government before HMG began negotiating seriously on the issues. On this interpretation, the British Government assisted in empowering activists: recognising the worth of their interstitial position and asking them to intercede on their behalf. According to this logic, HMG acknowledged its own relative weakness and created a further space in this open-ended process, another one within which the activists could play a political role. The British Consul, Alicante, is keen that Jeremy emphasise this is a campaigner's interpretation, not that of the Consulate.

These groups direct their main efforts at politicians. At the same time, they run press offices, to counter misrepresentation of the migrant demographic. They know that most English media, even the usually sympathetic *Guardian*, may stereotype Britons in Spain and France as jingoistic retirees (e.g. Engel 2016). For many anti-Brexit and rights campaigners, this is a caricature too far. BIE, at a continental level, dovetails this activity by working to correct simplifications with a website photo gallery which emphasises the range of British migrants. Its professionally taken images portray workers from diverse sectors, schoolchildren, single parents, entrepreneurs, pensioners, early retirees, families, students. Though most are lifestyle migrants, no one sub-group dominates.

Emoting, digitally, about the EU

Most of the campaign groups have multiple online presence: a closed Facebook group page, an open Facebook page, and Twitter accounts. If the first tends to be conversation between the converted, the last two are available to anyone who wishes to connect. Some campaign groups have also launched their own websites. It is above all these groups' Facebook pages and websites that are central tools and modes of practice. Both leaders and members recognise the closed-group page can provide a variety of supportive functions, several of them emotionally oriented. For them, the need is clear. The press has widely reported that Brexit has divided families and endangered friendships. Our interviewees and our personal experiences corroborate this. Siblings avoid contact; families split into factions; friends are cut. This has left some feeding their anxiety in solitude. Some find comfort communicating their condition online to the sympathetic. For instance, some members of the French-oriented group we will call FFU (France For Us) are explicit about the nurturing role their closed online discussion can serve for the Brexit-isolated and worried among their membership:

> Belonging to this group might help to keep us a little stronger.
> I have found this group an indispensable source of help and information and "solidarity" following the referendum result. Many of our

lives are so shaken to the core that it is simply not possible for us just to accept the situation.

We are angry, bereft and feel misplaced. . . . We are looking for information but also solace to get us through these dark days. We find the group and the contributions members make to it, invaluable.

I have found the views of FFU-ers to be helpful, because you know others out there feel the same, pithy, thoughtful, practical, positive.

At our Perigueux meeting, one France-based campaign representative reported that some of their members were spending several hours a day on their Facebook page. He said it provides them with 'a lifeline and measure of understanding they don't get with their families'. It was a 'force for positive change' for otherwise despondent anti-Brexiteers.

He went further: the Facebook page acted as 'a safe place for them to let off . . . about Brexit'; it was an 'outlet for feelings of grief, anger, fear and betrayal'. Some contributors to the FFU-page were well aware of this; they pinpointed the therapeutic function of the discussion, allowing them to vent strong feelings to co-believers, and then relax, knowing their exclamations will be condoned, unless too expletive or insulting:

For the last three weeks [mid-July 2016] I think we have all been in a form of political shock and horror I know I have and I have found myself in a state of flux regarding the life I enjoy in France and love. One of the symptoms of this emotional state is that we may have all felt and said things that are not representative of our normal behaviour.

[The FFU page], if nowt else good for letting off steam!

To me the FFU page . . . allows us to bitch about what we don't like about the whole situation – and let's face it, it can be very cathartic to have a good moan with like-minded people. Having moaned it clears the mind.

Mid-September 2016 the FFU-page moderator tried to modulate debate by reminding contributors, 'People, in general, are upset, frightened, and feeling insecure, so short fuses are to be expected'.

A new moderator, who took over in January 2017, pointed out to Jeremy that up until that date,

FFU was still very much in formation, without clear aims, and so was rather more chaotic in an emotional sense. That was in some ways deliberate, as it was felt to be important to allow members the space to express their feelings and let the group grow organically. . . . From January 2017 . . . we began to create a stronger identity, with FFU's specific aims as a container. The space to express emotions was still present, of course – and still is; the creation of community is a strong part of our existence.

In other words, the emotive potential of FFU's Facebook pages is a key resource for the organisation as a whole. Perhaps that is why, since 2017, applicants for admittance to its closed pages are carefully vetted 'to ensure they are "compatible" with the group ethos'. About half of the applications are turned down. Similarly, since late 2016, Bremain in Spain asks prospective members three questions and checks their profile. They uphold a 'strict Remainers only policy'; they want their closed pages to be a 'safe place' where participants can talk with 'like-minded people'.

Overall, the moderator plays a finely judged mediating role. He/she decides when to block further posts after a particular conversational thread appears over-extended: when the informative redundancy of new posts becomes high, and the overall returns are fast diminishing. He/she also reminds participants that trolls can always infiltrate their discussion group, that some messages appear to be posted simply to provoke reaction, and that upholding free speech demands that respondents treat others' views with respect, no matter how confrontational they might appear. The moderator tolerates mild profanity but quickly reminds those who edge beyond this of the need for equability. Catharsis should not turn into aggressive abuse; venting aggrievement must not smother the discussions' other functions. We have yet to see any posted criticisms of the moderator's judgements on any of these issues. Indeed, when debates have become heated, some contributors have noted the pleasingly tempered style of FFU's pages compared to the online discussions of some pro-Brexit websites. In the words of one, responding to an unhappy fellow member,

> A lot of the "leavers" did vote according to their consciences, as you say. I personally think a lot more did so because they simply wanted eastern European people out of Britain, like my father-in-law. They were not "imbeciles" of course they weren't, no-one on this group has suggested any such thing (to my knowledge). But if you go to the "Change Britain" Facebook page I think you will be shocked by the level of the bad language of the leavers on there – I certainly was.

These discussions on campaign Facebook pages also serve to keep members up to date about recent events, their potential implications for British citizens, and the reactions of group representatives to them. They are a way for representatives to keep up members' commitment and to take their reactions into account in coalition meetings. Indeed, members' testimonies provide rich empirical evidence informing reports that the campaign group has drawn up. The online discussion spaces allow members to propose changes themselves, and in the process make the pages more of a genuine forum where all members' voices, and not just the representatives', may be heard in an equitable manner. As one contributor put it,

> When did democracy become a one-way deal? Having a democracy means that we are allowed to challenge something we don't like. It's our right to try to change minds if we can. A forum like this has already encouraged me to come up with and submit a possible solution (see my post last week). Naïve it and I might be, I've never been a political animal, but I have been inspired to at least try. I think I have a solution for everyone. In the unlikely event that anyone takes notice of it – its origins are firmly rooted with the inspiration of this page and its contributors.

The view of FFU is that it, like other anti-Brexit groups, is striving via its online activity to engender in its members a sense of agency, to overcome any feelings of powerlessness. Besides encouraging members to participate in campaigns, the open FFU webpages keep them up to date on the latest stage in the negotiations on citizens' rights: the aim is to help them make the most informed life choices possible, e.g. to secure more firmly their rights to residence, or to apply for citizenship. FFU hopes its actions will have a cascade of consequence, its informed members in turn informing others of their rights, and so on. According to the FFU moderator, this 'not only helps reinforce community but also gives a sense of purpose and identity to people who are at risk of seeing both of those swept away by political circumstances'. FFU wants its members to feel they have some control over their lives instead of being forcibly carried along by events outside their command.

For similar ends, in 2018, Bremain in Spain began rolling out 'barnstorming' events, in association with the UK-based Best for Britain campaign group. The goal of these training sessions is to help Remainers manage conversations with Leavers more effectively and less emotionally. Feedback from participants revealed they felt supported by the training and contact with others. In the words of one Bremain activist, they 'are now finishing some conversations feeling they have made some good points, rather than being frustrated and more angry than when they began'.

Regaining self, re-stating home

In affective terms, activists' behaviour can be regarded as a means for them to regain an integral sense of self. A remarkable example here is Sue Wilson,

Chair of Bremain in Spain. In a 2019 confessional article, she revealed that before the referendum she was 'a political virgin':

> I felt perfectly content in Spain and in ignorance, and felt that any decision made in Westminster were of little consequence to me.

But the referendum changed her 'into a different person. It turned me into a campaigner'. She had heard that many Bremain members made similar statements, speaking of the 'damage' the referendum had 'done to their sense of security, their health and well-being, and the anxiety it causes about their future'. She noted her language had become stronger: a 'common side-effect among Bremain campaigners'. Brexit had taught her new skills and given her confidence, previously unimaginable, to speak publicly, even to crowds in the tens of thousands. By 2019, she could hold her own in political debate, contribute articles to the media, and give interviews on television and radio. In sum,

> I value one thing about Brexit: that it has given me a passion and commitment to change my future and, I hope, the future of others.
>
> (Wilson 2019)

Her example may be unusual, given the time she has committed to campaigning since 2016, but the point remains: agitating against Brexit can be a way to convert negative affect into a positive transformation of self, to build a new social network tied by a new set of common interests and energised by shared, strongly held passions. As the Bremain in Spain e-newsletter put it, in January 2020,

> We have all been through so many emotions in the past four years, none of them easy to deal with. We have cried, we have shouted, we have pleaded . . .
> . . . we must focus on the positives as much as possible. We've made new friends, we've learnt new skills, we've discovered new passions. Most important of all, we have learnt to appreciate our European home more than ever before. We may hate Brexit with a passion, but we cherish our lives in Spain – Brexit cannot change that.[4]

The sense of identity in play here has both individual and communitarian facets. Participants wish to retain or regain an integral sense of self. The campaign groups want to boost their individual members' sense of agency. Activists strive to build for all a sense of belonging to a new community, for migrants otherwise feeling isolated from their former social group of 'British'. The overall aim is to construct a new 'Remainer' community of activists and agents, secure in their purpose and reassured they are not alone. In the process, the more staunch of campaigners manage to turn personal despair into public zeal, assembling a renovated sense of self, and re-grounding what Wilson refers to as 'their sense of security'.

If we take identity as socially and personally central yet dynamic and contested, then the referendum is a rare example of what some activists class the 'theft' of their identity: a label they possess, and value has been 'stolen' from them. As one FFU-page contributor stated, 'I feel . . . cast adrift. I did not abandon my country'. Or, as a member of the Bremain Council put it to Jeremy,

> I was raised in Britain, worked in Germany, have lived in Spain for twelve years, and am going to Italy soon to marry my Italian boyfriend, where we will live. I feel very, very European. How can people take that from me?

Others complain they are being re-classed involuntarily from EU-citizens to migrants and third country nationals, while the campaign groups reject openly members' subaltern status and proclaim their collective actorhood by deploying their much-repeated slogan, 'We are not bargaining chips!' Turning activist can also be seen as a way to publicly declare opposition to this categorical dispossession, a means to retain a sense of their snatched identity. Dimitrios Theodossopoulos, studying the Greek financial crisis, saw locals' identification with communal indignation as an attempt to explain, and so subdue the crisis (Theodossopoulos 2013:208). Something similar is occurring in Spain and France. For both Greeks hobbled by austerity measures and British migrants fighting Brexit, their voiced indignation is empowering. Agency will out.

Some anti-Brexit migrants wish to go further in the clarification of their status. At the Perigueux meeting, when one representative spoke of 'If we have to go home', a good number of the audience immediately shouted, 'But we are home!' These expostulants seek to redefine the terms of the relevant political discourse. They wish both to separate natal citizenship from place of residence, and to retain a link between the two. In doing so, they aim to remind the British government and the European Commission that the conclusion of Brexit would not automatically lead to their mass migration back to the UK, nor to them all seeking to become naturalised French citizens. They wish to make the politicians realise that 'home' is as much defined by people as by legislators, and that any negotiated solution needs to take that into account.

For the activists themselves, as one stated, the social break caused by the referendum has been 'filled with fast, deep bonds with fellow campaigners'. It is difficult to avoid adding that some might also see prominent activism both as self-therapy, as the autobiographical article by Sue Wilson suggests, and as a means to create a publicly virtuous position for oneself. At the same time, they may be denigrated by some non-activists as 'keyboard warriors' (activists whose activity does not venture beyond their computer) or 'snowflakes' (delicate items which only endure in narrowly defined conditions). In a nice twist some Remainers have enacted a contemporary mode of symbolic inversion: taking on for themselves these derogatory terms, and the more offensive 'Remoaner', and revalorising them in the process. They work to draw their opponents' sting, to neutralise their intended slurs, by making them badges to appropriate with pride.

Of councillors and activists, morality and rights

Comparing the style of councillors with that of activists highlights what is distinctive about the campaigners and their modes of action.

British town-councillors laud an everyday ethics. Activists, though also migrant politicos, operate in a very different terrain. Like the municipal councillors, they have to bend to administrative procedure. Unlike them, they are entering much more extensive, amorphous, complex, multicultural spaces where appeals to a quotidian, vernacular morality, perhaps one more posited than tightly integrated or logically structured, hold little chance of sway. Instead, in the hope of achieving their aims, the contexts they are entering, and helping to re-fashion, force them to adopt a rights-based discourse.

A further contrast is their attitude to 'Europe'. Very few councillors referred to it as a relevant category. The topic was not even raised in Jeremy's 2015 interviews, and only in Fiona's when she asked specifically about it: councillors' horizons were limited mainly to the local, a bit to the regional, and rarely to the national; anything beyond seemed below their skyline. Studies of Britons in South West France in the mid-2000s and the early 2010s report the same (Drake and Collard 2008:227, Ferbrache and Yarwood 2015:79). In contrast, Janoschka, who fieldworked in the 2000s, found 'a strategic use' of European identity by foreign councillors as a foil to comment on local politics (Janoschka 2010:710). But this was most likely a consequence of the contemporary anti-expropriation campaign in his fieldsite, where protestors' concerted, repeated appeals to the EU were central to achieving their aims. Unlike the 'Euro-shy' councillors Drake, Collard, and we interviewed, anti-Brexit campaigners, by definition, make the EU and 'Europe' key frames within which they act. Their discourse would be incoherent without them.

The main priorities of the campaign groups are access to healthcare services; securing of pensions; unfettered travel; home ownership; exchange rates; votes for life; inheritance rules; choice of country of retirement. They couch their claims in a language of rights and citizenship. For instance, ECREU calls itself 'a lobby and self-help group set up to ensure that the issue of individual rights are foremost in the minds of those negotiating our future within the EU'; one of its two key aims is 'to protect the best interests of UK citizens living in the EU'. Similarly, the Governing Principle of the Alternative White Paper is: 'UK citizens currently resident in the EU and EU citizens currently resident in the UK should be expressly treated as continuing to have the same rights as they had before Brexit' (Golding and Morgan 2017:2). All the fundamental statements of the other campaign groups mirror ECREU's: there is no appeal to a common morality. Rather, as still members of the EU, they believe they have the right to claim and secure rights, and they argue within the terms of that discourse. In other words, the councillors wish to traverse a primarily ethical terrain to reach a just end and (in France) explicitly to embed themselves in local life, while the campaigners triangulate their course along the contours of rights-based discourse, i.e. along social, political, and legal lines.

Activists couch the possession of rights within a broad concept of citizenship. One highly innovative dimension of their campaign is that they are attempting to help generate a new style of citizenship. Traditionally, citizenship was defined within the frame of the nation-state (Isin and Turner 2002). The citizens of a country, and there were no citizens without countries, were those born and brought up within the borders of that country. This was the general rule, though of course there were always some exceptions: e.g. those born abroad to nationals working on government service or for colonialist end. This general assumption was so well-rooted and accepted that it usually went unquestioned. Thus, citizenship and nationality could mesh unproblematically, indeed almost invisibly.

The sustained development of the EU questioned that hitherto-assumed categorical rationality (Lister 2008). As it struggled towards the creation of a supranational entity, the link between citizenship and nationality was logically disrupted, and with it the assumed association of rights with nationality. For instance, Britons disgruntled about an aspect of governance became able, as EU members, to challenge the sovereignty of their own nation-state by appealing to a novel body: the European Court of Human Rights. Both the Court and the European Convention of Human Rights, which it upholds, are without precedent in English legal history and, as such, a particular bugbear for pro-Brexiteers. But logical disruption does not entail dissolution: though Eurocrats might wish to give further substance to notions of European citizenship, fiscal practice and actuarial processes across the

EU still 'presume a more or less static and bounded populace'; the fluid lives of migrants are confounded by fixed national laws, ones in need of change (Ackers and Dwyer 2004:463; Favell 2008). The sum result of these and related EU-phenomena is that EU citizenship remains a shifting, contested notion; it is that open contestation which activists exploit to create a dynamic space in which to develop their case. Indeed the legal theorist Patricia Mindus argues the best possibility of decoupling EU citizenship from its member-state counterparts, to the benefit of post-Brexit Britons abroad, is via a mass citizens' initiative (Mindus 2017:92–4). If this be right, the activists' power to set the agenda and steer the refiguring of EU citizenship is potentially all the greater.

The activists are not working towards 'shadow-nationality' (Ronkainen 2011), a 'citizenship of convenience', where migrants 'cherry-pick their rights and strategically evade certain territorially-defined duties' (Ferbrache and Yarwood 2015:82), or 'reciprocal rights', beloved of social contract theorists, which attend to both the rights of citizens and those of states; nor, unlike the claimants studied by Isin, are they usually claiming 'a right . . . to a liveable life when no such prior authorisation exists' (Isin 2013:32). Instead, they are advancing a maximalist position, seeking to pressure EU bodies and those of relevant member-states to guarantee as many of the rights they at present enjoy, and to guarantee certain rights they do not at present enjoy, e.g. votes for life. At the same time, some have pursued an alternative strategy: organising a petition, signed by almost 26,000, calling on the Spanish Government to grant dual nationality, a status which does not yet exist in Spain, for Britons resident in the country for over ten years.[5] In sum, as occurred in Italy, though there because of organised protest by underprivileged migrants (Oliveri 2012), activists can contribute significantly to open the boundaries of citizenship.

Perhaps a key difference between the councillors and the campaigners is that their initiatives have tended to be far more individual in style and rarely stretched beyond municipal boundaries. In contrast, from their very beginning, the campaigners have been united in their upset and anger. They share their indignation and turn it to productive end, ultimately to a continental boundary.

Bourdieu, Isin, habitus, enactment

Within practice theory, social life is enabled and reproduced thanks to habitus. Individuals can adapt, broaden, and presumably shrink their habitus, to a limited extent. Their degrees of freedom here are constrained. In contrast, Isin applies his approach to cases where change is central. He focusses on examples where radical challenge to the status quo and an open-ended process are at the core of relevant activities. It fits with an analysis of an innovative, ever-developing campaign in a novel situation. To use Isin's vocabulary, enfranchising non-citizen residents in Spain and France enabled Britons to enact EU citizenship in a novel manner, but within a stable context. The Brexit process is different. Even if the notion of habitus could be stretched productively to incorporate its multiplex dimensions, which we doubt, it would still fail to grasp in an adequate manner what makes Brexit unusual. For this is a future-oriented, geographically diverse process where only the general end is clear: British departure from the EU. Exactly how, on what grounds, is still being debated as we write in early summer 2020.

The anti-Brexit groups formed and grew very fast, by activists in France, who had some experience of campaigning, and those in Spain, who had little such experience. In Isin's terms, these campaign groups are examples of performative citizenship because they enact creatively rather than follow scripts (Isin 2017). They question subject positions and make claims for rights in the face of Brexit. As campaign groups evolve, grow, mature, transform, or decline they demonstrate new and flexible ways of being political, comprised of the people involved and the tools they use. The Internet enabled the level of organisation and speed of political response of the groups, and they appear to be having an effect in their dealings with different actors.

The sustained actions of the activists, via social media, in face-to-face meetings, in the streets, appears to have had, and to be having, a noticeable steer on the course of these debates and negotiations. Campaigners, as bottom-up agitators, are also challenging top-down arrangements. For they expose the otherwise seemingly monolithic EU polity as in reality a complex European institutional assemblage. They exploited this complex, multi-faceted phenomenon successfully as a variety of different stages where they can stake their claims and make their voices heard, as both securers and claimants of rights, who are operating provincially, nationally, and internationally.

There are still questions, however, which can be asked about the anti-Brexit organisations. For instance, the British Consulate, Alicante, has for some time queried how representative these groups are in the area. Of course, participation is to a great extent confined to those on the beneficial side of the digital divide. RIFT is transparent about this, highlighting who and where are most effected:

> While the RIFT Facebook Group and website are 'online' and by their very nature are accessed in the main by people who have internet access and use online groups and websites, there are an unknown number of British people who do not have either the IT knowledge or equipment

to make online applications. We have been made aware of this by our members who, often in the rural areas of France, know friends and neighbours who are in this situation, frequently older people.

(RIFT 2020:9)

But there are further worries whether the lopsided representativeness of these groups extends beyond that. Most of the migrants interviewed by Jeremy in the Alicantine interior had only a shady knowledge, if any, of these groups, though all were connected to the Internet. Moreover, even among migrants aware of the campaign not all wished to be agitators, however threatened their position. Finally, it has been long known that an important percentage of British citizens in Spain and France have energetically avoided contact with any branch of officialdom, let alone activism, as we will discuss in the next chapter.

Notes

1 For discussion of the origins of this group see www.thelocal.fr/20190103/how-a group of brits-took-up-a-struggle-for-millions-of-their-co-citizens-part-two (accessed 10 May 2020).
2 Currently, Britons residing abroad for more than 15 years lose their national electoral vote.
3 For further examples of digitally enabled advocacy groups effecting real change, see Postill (2012), Margetts (2013).
4 https://mail.google.com/mail/u/0/#search/bremain/FMfcgxwGDDjqhBprhXnBL zdBswpVlSNr (accessed 17 May 2020).
5 www.change.org/p/dual-nationality-for-brits-who-have-resided-in-spain-for-more-than-10-years (accessed 9 Jun 2020).

6 Rights and residency

The migrants we discuss have moved for the sake of jobs or a better quality of life, sometimes both. They are not fly-by-nighters but expect to stay at least a few years, maybe much longer, perhaps unto death. Some were dutiful, fulfilling the bureaucratic requirements of residing in a host country. Few worried or, as far as we could tell, thought about their entitlement to and enjoyment of a range of rights. The referendum changed these attitudes radically. It forced many to rethink their position on a range of central issues, and even to re-consider how they identified themselves. It also pushed some to act, and take their futures into their own hands.

In this chapter we chronicle and analyse these processes for our two field-areas. We do so primarily by investigating the different bureaucratic regimes of both countries at varying levels; migrants' approaches to official procedures, or studied avoidance of them; and their understanding of and attitudes towards varieties of citizenship. In these different ways, we here further our probe into migrants' agency and their calculated negotiating of officialdom, as they attempt to secure residency or gain citizenship. We conclude with comparative comments on relevant Hispano-French differences and similarities in these spheres of action. While the previous chapter focussed on the actions of migrants gathered into campaign groups fighting a cause, this concentrates more on migrants as individuals trying to traverse an, at times, dark labyrinth.

Spain: to encounter a foreign bureaucracy

'Spain is different', the country's government used to tout from the late 1950s onwards, as it stimulated the shift from a stagnant agricultural economy to the advent of mass tourism. That slogan no longer holds good for much of the country, today an industrialised state with a very developed tertiary sector, and administered by a modern civil service. However, the meeting of bureaucrats and migrants is not always positive, especially if the Britons have not integrated well into the country.

Britons who migrate to Spain are required, after residing there for an initial three months, to register with a local station of the National Police or the provincial branch of the Office of Foreign Nationals, who grant them residency permits. An incomer who registers as a resident pays income tax to the Spanish Government. On application, they have to declare their worldwide assets: this does not influence their tax rate; the authorities only wish to know what a resident can cash in if the need arises. Possessing a residency permit is necessary in order to access healthcare services for free; even then it is only available for the salaried or those registered as self-employed. New migrants have also to add themselves to the *padrón* (town-hall register) of the municipality within which they reside. Being *empadronado* is necessary in order to buy a car or a house; it also allows EU citizens to vote in and stand for local elections, but they have to declare formally their intention to vote and have to register on the electoral roll. To remain registered and be able to vote, they have also to confirm their padrón status every two to five years, depending on a town-hall's interpretation of the regulations. On top of that, residents who are empadronado are liable for council tax. If they change residence, they have to re-register on the padrón of the municipality they have moved into. Immigrants who can demonstrate they have been 'continuously resident' for at least five years can apply for permanent residence.

A Briton who has been in the country for three months and intends to remain, commits a crime, *un crimen*, if they do not register with the National Police. To be empadronado is also a legal requirement. However, neither Jeremy nor the British Consulate, Alicante, have yet to hear of any migrant, whether British or Continental, being charged, let alone taken to court for failing to do either. It appears these are EU regulations the forces of the Spanish state, from the national to the municipal, choose not to implement. Further, in Spanish law, a Briton who divides their time equally between each country is considered as occupying a very grey area. According to consular sources, the Spanish Government, on drawing up this legislation, had not foreseen that migrants might spend, each year, six months in the UK and six months in Spain. Indeed, the British Consulate, Alicante, advises Britons not to register if each year they spend seven months in the UK and the remainder in Spain: otherwise, they would have to re-register every time they returned to Spain. In the opinion of the Consulate, the Spanish Government is 'not very user-friendly' in the process of becoming a resident: its website, for instance, is only in Spanish.

This apparent lack of bureaucratic interest in migrants is mirrored by the behaviour of those incomers who minimise contact with any branch of

officialdom. Their reasons are various. Some fear coping with an administration they do not understand, which conducts its business in a language
most of them do not speak. Thus, they steer shy of electoral rolls, police
registers, and anything else which smacks of bureaucratic monitoring. Others, who do attempt to engage, found they were given conflicting advice by
post-holders and bureaucrats. One said the mayor of their municipality told
them they did not 'need to bother' registering on the padrón. Some, unaware
that Spanish town-halls had significant power and autonomy, were confused
by municipalities implementing regulations in different ways. For instance,
one south Alicantine town-hall would only allow migrants to register on the
padrón if they had already obtained a residency permit. As far as the British
Consulate was aware, this was the only municipality in Alicante which had
so far required this. O'Reilly researching in the early 2000s found a similarly
confusing diversity of regulatory implementation in municipalities along the
Costa del Sol (O'Reilly 2007:287). Others said they found the business of
re-registering on the padrón or of re-declaring their wish to vote, every so
many years, tiresome and forgettable, so they did not bother. This is one
reason why the percentage of enfranchised foreign residents who vote in
municipal elections is disproportionately low. One migrant colleague spoke
to Jeremy of being given 'the usual round around', as though it were a to-
be-expected occurrence, when trying to fulfil a bureaucratic requirement:
she was obliged to visit a series of offices in Alicante city, before finally
having her papers signed back in the first department she had visited. She
also referred to

> a *funcionario* (public servant) mindset here, because until recently a
> public service post was considered to be a prestigious kind of employ-
> ment as it guaranteed workplace security. I have had a few run-ins with
> public servants who adopt this mindset, when I was actively working
> or teaching on the campus of [X University], and have even been left
> waiting for a couple of hours while said funcionario was away at *alm-
> uerzo* (lunch).

Further some, such as those who had tried to register new political parties,
had found bureaucrats put obstacles in their way or were generally unhelp-
ful. Another complained of what they considered 'unfair' procedures for
applications to become Spanish citizens: there was only one day a week
foreigners could attend the relevant Government office, and they had to line
up, while a Spaniard acting on behalf of an applicant could 'jump the queue'.
Some, aware of illegitimate practice, were wary of contact with officialdom:
four spoke of new arrivals in the 2000s who had felt forced, despite their best
intentions, to make private payments to councillors for securing municipal

permission to build their home on the plot of land they had bought. These interviewees added that these practices appeared to have ended.

And yet, when in April 2020, Jeremy block-emailed his former interviewees, requesting examples of unhappy experiences with Spanish bureaucracy, eight replied that he was in danger of perpetuating outsiders' negative stereotype of local ways. Jeremy accepted this mild admonishment and block-emailed a more balanced request. Respondents, who had all lived in the country for over a decade, emphasised the need for incomers to learn the language, accept cultural difference, and practice patience. They underlined the centrality of courtesy, calmness, and a desire to engage with the locals and their ways. Some also underlined the sustained wish of Spanish bureaucrats to accommodate to British migrants, even the monolinguals. In the words of one, 'I can't say how much Spanish people bend over backwards for the Brits'. Indeed, in a pointed Anglo-Spanish comparison, one rubbished British bureaucracy as 'inept, obstructive, and difficult'; another detailed how he had left the UK because of hometown bureaucrats. Some migrants with money but not the linguistic confidence paid *gestores* – 'managers', hired intermediaries between their clients and Spanish bureaucrats – as these experts in administrative realities know 'how far you can push'.

All these respondents were united in their criticism of Britons who, as they put it, expected Spain to be but a sunny version of the UK, with cheaper beer. One chided those who did not learn the language, nor have the right 'mindset'. According to her, these British monoculturals would not prepare the requisite paperwork before meetings with officials, in which they might become 'quickly abusive'; they could also lose their temper too easily, for example when told their car has failed its MOT (ITV in Spain) because of an obvious detail. Another emailed of

> The ones who sail into the Town Hall and shout in *English* at the staff. I have met and helped many such people. They are the last bastions of the British Empire, ordering the natives around . . .
>
> The neighbour opposite, saying she did not know there were any planning rules in Spain (yes she did say that) built a high wall and had to take it down again. She was so incensed that she sold the property and left Spain. Seriously!! (Original underlining).

According to one, these incomers

> expect Spanish authorities to speak English. . . . [Yet] I have heard several of these people say they came to Spain to avoid the overwhelming presence of foreigners in England, often paradoxically complaining that these don't speak English.

In sum, Jeremy's respondents made patent their lack of sympathy for British immigrants who remained within Anglophone enclaves and kept up caricatures of the local life beyond their monolingual redoubt. Spanish bureaucracy may be daunting and off-putting to the cross-culturally inexperienced, and even occasionally obstructive to more integrated migrants. But the judgements made by the culturally unconnected cannot be taken as indicative of the whole. Moreover, as some underlined, occasionally Kafkaesque procedures and practices of officialdom could also be encountered in other European countries, such as the UK.

Spain: to register, or not to register; to reinvent

The lack of interest in or fear of Spanish bureaucracy evinced by some has enabled something consular officials and local administrators have long known: that, until very recently, many Britons residing in Spain were unregistered. Their overall number cannot be specified with a satisfying degree of exactitude, despite repeated attempts by consular officials, Spanish civil and municipal servants, and social scientists to guesstimate the size of this population. Maybe they constitute 50% of the Britons in the country, maybe more (O'Reilly 2017:141). They are, if you will, the dark side of lifestyle migration.

In 2019, Jeremy did manage to speak with several 'unregistered'. But these were almost all residents in the interior. Despite his repeated efforts, he failed to make any contact within the coastal caravan parks, where the unregistered were said to be concentrated. All of his interviewees stated that they had paid, or did pay their taxes in the UK, like any other law-abiding British citizen; the reason they did not register in Spain was not part of a generalised strategy to avoid officials, but the more specific one of evading those Spanish taxes which are higher than their UK equivalents. Depending on one's range of assets and mode of income, Britons may well pay more to the Spanish state than they would back home. One resident British lawyer told Jeremy, 'If someone wants to know whether to register, I ask them their income and where it comes from. Then we decide'. Several said to him, whether you chose to register 'depends on your accountant to some degree'. For instance, inheritance tax is much higher, and the law dictates what percentage must go to which family members. One interviewee exploited the regulation that only one member of a couple needed to obtain a residency permit. Though he needed long-term healthcare, he had never registered, as his spouse had already done so. Another, who worked informally, said that self-employed foreigners who had registered were obliged to pay 300 euros a month, for their '*seguridad social*', to fund their free access to healthcare. This interviewee did not wish to pay 3,600 euros a year: it would make their relatively meagre locally earned income unprofitable. Jeremy spoke with both Leavers and Remainers who were unregistered: in every case, their rationale was predominantly financial, and independent of their attitude to Brexit.

All of the unregistered interviewed were openly sceptical about the value and effectiveness of the campaign groups, discussed in the previous chapter. For those opposed to Brexit, their scepticism justified their lack of participation. An activist who read a draft of this chapter stated these attitudes did not surprise: for her, the anti-Brexit campaign was a righteous fight; and it is hard to be righteous 'when hiding from the authorities in order to dodge taxes'. In her experience the unregistered tended to exhibit an unassailable sense of British superiority: no foreign government would be so 'daft' as to dare to throw an English person out of their country. These feelings of nationalist superiority dovetail with the anti-immigrant dimension of the Brexit bloc, i.e. already indisposing them towards any anti-Brexit campaign. As another stated,

> It certainly appears that some don't see why the Spanish authorities have any jurisdiction, because "they are British" . . .
>
> There are those who have left their mobile homes (on campsites, all unregistered. . . . The lawyers say they wanted to go before they would be "caught out" as illegal, or that they were so disgusted that they would not be allowed to come and go as they wanted.

Some of the 'unregistered' interviewed by Jeremy present their attitudes to the future as pragmatic: if the result of the Brexit negotiations allows them to remain, they will; if the results are otherwise, they will fall back into the arms of the British state, however welcoming or not at that time of return. For the British Consul, among many others, the worry was that, come the final separation of the UK and the EU, these individuals would become in effect, what she and Jeremy termed, 'status-less' citizens. If they become in sudden need, it is very unclear what services they could seek from Spanish authorities, and what rights they might retain in their country of origin, the UK. As of 2019, though many unregistered were signing up (lawyers spoke of a 600% increase), some had decided to wait and see what the final result was. Consular officials thought this procrastination 'very sensible'.

The unregistered keep away from the National Police for financial reasons. They strive to take full advantage of beneficial disparities between British and Spanish tax regimes. In contrast, the reinventors, whom we discussed in Chapter 2, are creating a past to suit their present, a task only limited by their imagination and persuasiveness. Many of Jeremy's interviewees, whether Leavers or Remainers, are relatively distinct from both these groups. Like the unregistered, they are concerned with personal budgeting, but their ultimate aims are politico-moral, not monetary. They are more interested in grand questions of identity and governance than the minutiae of domestic spreadsheets. Like the reinventors, they pursue a dream, but in their case one of collective destiny, not individual biography. They put a socially grounded ethics before self-aggrandisement. Both the Leavers and Remainers interviewed uphold principles which cut deep; to them, the reinventors are an amusement on the side; and for the registered among them, the unregistered are a superficial fact of daily life. If anything, they are to be mocked for their penny-pinching evasion of the state, and the precarious situation they now find themselves in: they 'wished to hide away; but now are having to do all the things they avoided'. Their critics, who had long ago bothered to learn to cope with Spanish bureaucracy, stated, some with satisfaction, that though the unregistered had been saying since the referendum 'Nothing's going to change', from January 2019 many had started to panic, were now visiting solicitors for advice, and making their first steps towards registration. 'People are running around like blue-arse flies trying to get up to date'.

For most of the Leavers and Remainers Jeremy interviewed, the unregistered put money before principles; they judge dedicated reinventors indulgent make-believers, a cause for a wry smile, little more. But for those interviewees engaged with the pros and cons of Brexit, Britain's relation with the EU is not for trivialising; it is a phenomenon of a different, core order: a moral one.

Spain: to be a citizen

One way to secure lifelong residency in the country is to become a Spanish citizen. But there are layers of ambiguity here. First, many Spaniards and migrants speak of '*nacionalidad*' ('nationality') and *ciudadanía* ('citizenship') as though they were the same. The official Spanish Government website concurs: 'Both terms are practically synonymous'.[1] Second, there is some confusion about the regulations surrounding Spanish citizenship and their implementation. The Spanish Government only grants dual nationality to those from a limited number of countries, mainly Latin-American ones: Great Britain is not currently included, though anti-Brexit activists have long campaigned for its inclusion (e.g. Tremlett and Chislett n.d.; Macbeth 2019). Applicants from the UK are supposed to renounce their British citizenship though, as journalists have claimed, 'in practice many keep their British passports, as they do not have to be given up' (Collinson and Kollewe 2019). The Bremain in Spain website hints at this strategy:

> So long as you are Spanish whilst in Spain, and do not have a problem with denouncing your British citizenship to the Spanish authorities,

then you need not relinquish your British passport or citizenship – at least as far as the British authorities are concerned.

(Bremain in Spain 2020)

Similarly O'Reilly states that one does not have to give up a British passport but that, in Spain, 'you are only ever considered a Spanish citizen, not a dual citizen' (O'Reilly 2020:30). Even the Real Instituto Elcano, a semi-autonomous think-tank funded by the Spanish Government, admits 'citizens of almost all EU countries that acquire Spanish nationality *supposedly* lose their own' (González Enríquez 2014, our italics. Also Rodríguez 2013:12). According to the British Consulate, Alicante, the attitude of the Spanish Government appeared to be: if you don't inform us, we don't mind. On top of that, for the privilege of formally, physically handing over one's passport, the British authorities, as of 2018, charge £372.

British migrants might be ready to register, especially since the commencement of Brexit. But even those committed to a long stay appear unattracted to gaining Spanish citizenship. Though O'Reilly states that 'more and more [Britons] are applying for Spanish citizenship' (O'Reilly 2020:29), their numbers remain low: only 54 acquired citizenship in 2017, just 56 in 2018 (see Table 6.1 page 99).

France: a language of rights and citizenship

French Government regulations make no provision for the compulsory registration of Britons wishing to stay. While non-EU migrants (third country nationals) must apply for residency permits through their departmental prefecture, EU citizens, since 2003, are exempt from this system. EU citizens can make voluntary applications for a *carte de séjour*, covering the first five years of residency; or a *carte de resident*, for those with more than five years of residency (see Service-Public.fr 2020). But Fiona is aware of very few Britons who obtained them. It is more common to hear of Britons whose voluntary visits to the prefecture were met by discouraging bureaucrats. As one explained to Fiona about arriving in France in the early 2000s 'They just shrugged at me and shook their heads insisting I didn't need any residency card'.

Although there is no systematic means of registering one's stay in France, Britons are required to comply with other bureaucratic state norms. These include local taxes for property owners; declaration of capital gains and wealth; income tax for those staying in France indefinitely; healthcare cover for stays of over three months; business registration; car registration, electoral voting. To stay in France permanently or indefinitely, i.e. longer than three months, Britons are required to align themselves with these nationalised norms. Some Britons were surprised: they had taken 'free movement' quite literally, as something 'unfettered' where rights they already enjoyed would extend in an unbounded manner across member states. One couple expected British forms of bureaucracy to be reproduced in France:

> We basically didn't know how the system works. You assume that everything works on an English basis around Europe, it doesn't . . . It's a different country.

Britons interviewed by Fiona in 2008–09 did not talk about residency permits but mostly explained how they legitimised their moves in other ways; for example, 'We chose to be resident and domiciled in France. We've changed everything; tax, social security, insurance, car registration, banks, health care'. Most interviewees corroborated, revealing how they had fulfilled various bureaucratic conditions by completing forms; translating official English documents (birth, marriage, death certificates) into French; booking and attending meetings with different bureaucrats (at the mairie, tax office, prefecture, health insurance office, social security department, chamber of commerce, etc.) One Briton described the process as 'part of the mechanics of trying to live in two places. . . . The rules are there, and you just have to go with that'. Many described French bureaucracy as 'unwieldy', 'complicated', 'confusing', and the process of assimilating into it 'slow', 'time-wasting', 'repetitive'. Many Britons found themselves repeatedly visiting the same offices because their dossiers were said to be incomplete, there were several stages to each process and, as in Spain, they often reported passing through various offices before returning to their start point. These experiences became normalised perceptions of French administration among the British. They were not peculiar to them: Fiona heard similar descriptors from French acquaintances.

Britons discussing registration often did so in terms of morality. One said, 'As soon as you enter the country you're meant to be registered for tax . . . you have to be registered with the right people'. A couple told Fiona, 'You're in somebody else's country; you're not in your own so you've got to do it their way'; similarly, 'If you come here then you do as the Romans do and you pay in the system'. Expressing themselves as good people behaving correctly, some critiqued their fellow nationals who failed to comply: 'You don't just come here and take all the good things like the cheap wine, lovely weather, and nice food'. These were France's 'unregistered' according to compliant Britons.

Some Britons told Fiona that they did not fully comply with French bureaucracy. A migrant's access to property, healthcare, or establishment of a business in France is not reliant on evidence of residency. Rather, these systems function relatively independently. Many require migrants to declare themselves, and this can give rise to partial or selective registration. For instance, some Britons relied on the European Health Card for access to healthcare, earned money cash-in-hand, and survived easily enough without formally associating themselves with French bureaucracy. At the extreme, it meant that they were almost unknown to agents of the French state, having either overlooked certain administrative procedures or managed to avoid them. In the latter case, one revealed:

> Even though I have not contributed to the system (income tax and social security), I have not been a burden to French state either, I don't claim medical or anything like that. If I use the medical I pay out of my pocket.

For most of the deliberate cases the motive appeared to be potential financial gain. As one couple said, 'We fudge it' because 'taxes, social security, cost more than they do in the UK'.

Strategising can bring benefit but also pose risks. One Briton explained how her husband had had to return to the UK with a broken leg because they had not bothered to register with the French health service. Another who described himself as 'working under the radar' said, 'I want to put it right. But if I am caught working now, there could be serious penalties'. Someone in a similar position explained his reliance on word-of-mouth on the grounds that advertising could expose him to authorities. He remained worried about the future, 'What would happen if the grapevine fails to work?'

Some Britons dividing their time between France and the UK (and sometimes a third country), described their situation as 'confusing'. These were not necessarily people who overlooked administrative procedures intentionally; they just had not learnt how exactly these regulations applied to them. One pair of interviewees presented themselves as coasting in some middleground of unknowing: 'We've been here three and a half years, and we still pay our tax in the UK. But it's not quite right'. Another couple said, 'we thought we were doing everything right with finances and tax then three years down the line someone told us different and we thought, "Oh no"'.

In sum: whether deliberately or accidentally unregistered, Britons seemed to get by. Fiona knew of only three interventions from the authorities (relating to tax, health cover, and driving licence). Like citizenship, which is barely questioned until it ceases to function as expected, bureaucratic life seemed to play out in much the same way for these Britons. This situation may well have continued largely unquestioned if it had not been for the 2016 referendum result. Quite suddenly, many Britons in France became very conscious of their lack of official status: how to prove it, whether they needed to, whether they were even capable of doing so. Some, unknowingly, have fallen foul of EU requirements for lawful residence.[2] For many Britons, substantiating residency became important post-referendum.

France: registering as resident

Voluntary applications for residency permits rose dramatically following the referendum. In 2015, 189 applied; the next year, 1,319; and in the first eight months of 2017, 1,852. This translates as a rise, between 2015 and 2016, of almost 600% (Benton et al. 2018:17).

Britons told Fiona they applied for residency permits in efforts to ward off a deepening sense of precariousness about their status in France. In the summer of 2016, one man spoke on behalf of his family, 'We will have to apply for a residence permit to stay here'. Another reasoned, 'I think I need to get a carte de séjour'. A third said, 'It makes the future uncertain to have no status. I'm applying for residency now'. Their plans were driven by wanting 'to show we're legal', 'to show we're here correctly', 'to have documentation'. One Briton explained, 'If anyone questions [my right to be here], we have this document . . . It's reassuring to have some paper that shows I can enter the country. I cross the border a lot'. Another likened his 'shaky ground' status to his understanding of the 2018 Windrush Scandal in the UK, where British-born Afro-Caribbeans were wrongly denied legal rights and, in some cases, deported. One woman was explicit: she felt 'extremely vulnerable . . . because of what's happening with the Windrush people'.

Anxious Britons booked meetings with bureaucrats at their local prefecture; completed formal residency application forms; sought and translated legal certificates; engaged translators; collated utility bills, travel documents and bank statements. They created a dossier that could be presented to prove lawful residency, according to guidelines produced by the French government. Some Britons decided for themselves to perform these unwanted tasks; others were spurred on by word of mouth: 'We heard that the neighbours got theirs, and how easy it was. So we've done the same'. All this happened before any formal advice was given from the French or UK government for Britons to take such actions. That advice did not emerge until late 2018, when Britons were advised to apply for residency permits as temporary measures, which would clarify legal residence ahead of a new registration system for British citizens.[3]

Many Britons known to Fiona acquired some form of residency permit successfully. Fewer did not. Articles in the media tend to imply the opposite. Experiences varied, however, from straightforward to more complex, differing between departmental prefectures and shaped by personal circumstances. In Fiona's case she queued for an initial meeting to collect a hardcopy form for completion and a list of documentation required to pursue her application. She was assigned a date to return to present her dossier. In the latter meeting, all paperwork was checked, photocopies taken, and a temporary permit issued. The formal permit arrived several weeks later. Elsewhere things did not run as smoothly. A prefecture bureaucrat told one of Fiona's interviewees in early 2017 that he did not need a permit as an EU citizen. Another Briton who had encountered the same elsewhere, around the same time, grumbled, 'If I were German, I'd have no problem getting my carte de séjour, but because I'm British . . . ?' Benson noted similar administrative responses from her own research in the country:

> Even until late 2018, local bureaucrats were dissuading British citizens from applying for these permits, uncertain about whether, as soon-to-be non-EU citizens, these Britons were even eligible for these.
>
> (Benson 2019:3–4)

She notes further that prefectures were struggling to cope with a high volume of applications (Benson 2020). One Briton told Fiona they had spent several weeks in late 2018 and early 2019 trying to book an appointment with the prefecture only to have it cancelled a week before the scheduled meeting, due to 'the department claiming they would not be issuing any more permits'. A year later the situation had not been resolved. In other examples, Britons reported being 'misinformed' by bureaucrats, receiving conflicting advice, and providing seemingly necessary documents that 'weren't even looked at', including French translations of British certificates, which had incurred fees into the hundreds of pounds.

In sum, it is patent the administrative practices of prefecture bureaucrats were not uniform or consistent and Britons had different experiences depending where they lived. This advice could be strikingly individual. Friends counselled one couple, interviewed in 2019, to try to see 'the woman rather than the man' dealing with applications in their prefecture because she was 'more friendly' and more likely to grant the permit. Such stories did little to reassure the already concerned migrants rather, they confounded their felt sense of precarity and liminality. Their futures appeared dependent on the personality of bureaucrats who had the power to grant or deny permits.

In addition to bureaucratic variations, some struggle to meet the conditions to live in France, or to prove their residency. One couple told Fiona about an elderly neighbour whose only income was a UK state pension which would not pass the financial means test for a residency permit. Another Briton was self-employed with an irregular monthly income. A third fell outside the application categories (worker, student, retiree, economically inactive) because he was employed in the UK. He told Fiona, 'Even the British Consulate couldn't advise me where I fit'. He found himself having to register as 'economically inactive' and conforming 'to a set of requirements that were probably set up for rich folk'. This is but a small sample of varied experiences and inequal access to residency permits due to personal circumstances relating to issues such as income, employment, health, disability, ethnicity, family situation, dependents, separation and divorce, and widowhood (Benton et al. 2018; Benson 2019, 2020; RIFT 2020). Critically, many had thought themselves legal or legal enough, and this was now being formally questioned.

Moreover, successful applications did not necessarily forestall individuals' concerns for the future. One confused Briton pointed out the date on a newly acquired residency card: 'It's valid from the date I signed the form at the prefecture, and that's what, four years after I arrived'. Another received a permit 'only valid for one year': the process 'and anxiety' would have to be repeated in another 12 months. A couple in one location with more than ten years of residency were only issued with a permit valued for five years. It is difficult to determine accurately why anticipated outcomes were not fulfilled, but these examples show that justifying residency in hindsight can be difficult. These temporary measures can leave these Britons feeling anxious and concerned about long-term residency. Some thus looked to citizenship.

France: citizenship

Another way to secure long-term residency is to claim citizenship: French citizenship or that of a third country which will re-secure EU citizenship. British citizens acquiring French citizenship increased from 374 in 2015, to 517 in 2016, and more than doubled the following year, to 1,733. They rose a further 89% in 2018 (see Table 6.1 page 99).

Principal reasons Britons gave for considering and actually applying for French citizenship post-referendum included securing the right to remain in France, and to the broader entitlements bestowed by EU citizenship. One Briton told Fiona, 'I decided my life is here in France and I applied for naturalisation at the start of 2017'. Others revealed that they were 'concerned about being able to stay'. Some went on to express that they were considering citizenship 'to guard our rights', 'So our rights won't be removed', and as noted by one young woman, 'I ultimately expect to be unable to work or live here in the future as a UK citizen'. A couple expressed the different opportunities citizenship would bring to them and to their children:

> We are strongly considering getting French nationality for ourselves and our children so they can move where they want to, and we don't have problems staying here.

Birkvad's (2019) discussion of citizenship as a form of legal standing guarding individuals against 'liminal legality' helps us to understand this rise in British interest in citizenship. Uncertainty, confusion and lack of clarity about the future status of Britons in France, as well as lack of documentation confirming legal residency, has meant that involuntary removal of EU citizenship has created an anticipated gap in access to rights. This highlights their liminal status. For example, one Briton expressed that 'Brexit exposed me to vulnerability and risk', including ongoing access to vital healthcare, and ultimately an ability to remain 'at home' in France. The euphoria he felt at being granted citizenship was clear in his exuberant expression of 'Brexit can't affect me now'. Becoming a French citizen overcame his sense of liminal legality. Others expressed how 'We'd have never considered taking French citizenship before, it wasn't needed', making it clear that anticipated removal of rights was a key motivation.

There were also instances where Brexit was a trigger to carry out an existing intention to naturalise. One woman expressed, 'We were going to do it anyway'; another, 'We've had the paperwork for years, just haven't got around to it'. As was the case with bureaucratic reaction to residency applications, experiences could be very individual: Adam Thorpe, who has lived more than 20 years in a Cévennes village, described its former maire, as an '*ultra-gauche* detester of all incomers . . . (who) "lost" our naturalisation papers for a year'. Gaining dual citizenship took his wife and him 'four years of patience and paperwork' (Thorpe 2018:196, 199).

Despite Brexit being a catalyst for some citizenship acquisitions, deciding to apply for naturalisation can be complex. It is not always a migrant's first choice. Applying for citizenship was expressed by several as a 'backup' where they 'could always take out French citizenship if staying becomes an issue'. Some felt forced towards citizenship, as one couple stated, 'We're concerned that we might need to take up French citizenship if we decided to remain'. Another pair explained applying for citizenship 'only if we have to'. These examples convey a tone of necessity rather than optional action, but something that these Britons would be prepared to do to be able to stay put.

Citizenship applications can be made after five years of unbroken residency and require justification of resources, healthcare, language ability and integration (Service-Public.fr 2020).[4] These requirements mean that citizenship is considered outside the scope of possibility for some Britons. A widow in her seventies told Fiona, 'I don't think I'd ever pass the language test'.[5] Quite a few had not yet resided in France the necessary five years. Delays to the Brexit negotiation period were welcomed by some for this reason. Those who had been in France for five years also expressed reservations because they 'had no proof' or legal documentation of their residency. Others were worried that applying for citizenship would expose them to scrutiny from the authorities and hinder their ability to remain. In January 2020 one such example was the topic of media coverage in which a long-term resident (also with two children in France) had been declined citizenship because he could not prove a regular and sustained income of sufficient value (*The Local* 2020). This issue was expressed to Fiona several times: one woman stated that citizenship was 'high risk. If it fails you are out on your ear after 25 or 30 years'.

These examples demonstrate a post-referendum shift in dialogue as Britons actively discuss and consider naturalisation. Some consider citizenship as a safeguard or stabiliser to their current and future status as Britons in France, others express it as a necessity, yet others as an obstacle or barrier that may not be surpassable. In a study of post-referendum understandings of Irish citizenship, Wood and Gilmartin highlight the nature of strategic planning where 'National identity is here seen as a fluid category, to be accessed strategically in order to facilitate other lifestyle choices' (Wood and Gilmartin 2018:231). This appears to be the case for many of the examples discussed earlier. This pragmatic approach was evidenced further in alternative pathways to citizenship: in external forms of citizenship.

External citizens

External citizenship is held by those who do not live in the same country as the passport they hold. It arises when individuals seek to acquire citizenship from a third country, often for pragmatic reasons (Harpaz and Mateos 2019; Joppke 2019). The term 'instrumental' is often used in conjunction with external citizenship to emphasise its potential benefits, e.g. 'as an insurance policy' (Joppke 2019:860) or as 'asset building, an exit strategy, risk diversification, intergenerational wealth transmission, family protection, welfare benefits, increased opportunities, tax avoidance and other uses' (Harpaz and Mateos 2019:849). Acquisition of external citizenship is apparent among a few of the Britons whom Fiona interviewed. Four of them had applied for Irish citizenship after June 2016, and one couple had acquired Cypriot citizenship.

Those acquiring Irish citizenship made it very clear that they did so to retain their EU status, which simultaneously secured their continued right to reside in France as EU citizens. One new Irish citizen had selected Irish over French citizenship as the former is based on ancestry which he claimed was 'easier to prove' than French residency. A second person had not yet lived in France the necessary five years, while a third was uncertain that she would pass the language test and interview. The fourth individual captured the complexity of citizenship decisions by adding, to the reasons earlier, that he was 'proud' to claim his 'Irish heritage'. Asked whether these indi-viduals would also apply for French citizenship in the future, they suggested it would not be necessary so long as Ireland and France remained part of the EU (a claim that one Briton made to Fiona about the UK in 2009).

In a further example, one couple had acquired, in their words, 'the right to remain in France' by purchasing Cypriot citizenship. The country's invest-ment scheme for naturalisation allows individuals to purchase national citi-zenship through a minimum investment of roughly two million euros (see Ministry of Interior 2018). This simultaneously bestows EU citizenship that enabled this couple to continue living in France under the principles of free movement: as they stated, 'It's no problem now we're Cypriot'. Fiona heard of another couple who had acquired citizenship in this same way when their application for French citizenship had been turned down.[6] Given the high capital costs required to invest in such citizenship schemes, this pathway to secure EU citizenship and continue living in France is exceptional and unlikely to be a realistic option for most Britons living in France.

Towards a Hispano French comparison

Spain and France are fellow members of the EU. Yet they implement EU legislation in different ways, which it is up to their respective cadre of bureaucrats to interpret in their own manner. These cumulative differences have important consequences for Britons residing in their countries.

In Alicante it was only from the start of the referendum debate in the early 2010s that Jeremy heard Britons begin to speak of 'rights' and 'citizenship', whether British, Spanish, or of the EU. Fiona, during her 2008–09 fieldwork, noticed 'that Britons assumed, rather than failed to recognise, their supranational connection' (Ferbrache and Yarwood 2015:77). Very few acknowledged their status as EU citizens or the rights that entailed. Even when she prompted discussion, interviewees rarely articulated a language of citizenship or of rights (see also Drake and Col-lard 2008; Benson 2011). In the days following the referendum she began to overhear Britons, at cafes, in the market, around dining tables, using a new vocabulary of 'rights' and 'citizenship'. This was corroborated by those she spoke with, interviewed, and by online social media, including the emerging campaign groups. This language was at the forefront of con-versations relating to Brexit and Britons' futures in France.

British migrants in both countries might have started talking in the same way, but their national contexts were dissimilar. For a generative difference between the two states was that Spain had a formalised system of registra-tion, which it did not bother to enforce, while France did not even have clear procedures for registration, and its local bureaucrats were often unsure if any registration was necessary. While 'the unregistered' exist in both countries, the term means different things in each one. Moreover, unregistered status is co-constructed by migrants as well as *de jure* and *de facto* forms of state legislation. This difference between Spain and France had an important impact on the sources, and sense of anxiety and uncertainty some Britons experienced after the referendum.[7] In Spain, Britons who had chosen not to register either left or knew what paperwork they have to do: some are delaying registration as long as possible. In France, the path to registration is less obvious and successful application much less certain. Thus some Britons in France, unlike their counterparts in Spain, have been told they lack necessary documents or income, and their continued residence is under threat; they have become British citizens only, in effect mired in a transitional stage between their terminal status as EU citizens and that of third country nationals, while their future position is negotiated.[8]

In a similar manner, acquiring citizenship is not identical in the two countries. Both require candidates to take a language test, but the necessary level is higher for the French test (B1) than for the one in Spanish (A2). The exam in knowledge of French culture comprises 60 questions, to be answered in 45 minutes, from initial queries about candidates' motives for applying, to ones on the identity, political system, history, and culture of the country. The Spanish equivalent has 25 questions, to be answered in the same time, which cover a comparable but much shorter range of general questions. The last one for the 2020 model exam is,

On a menu, an almond tart is

a)   a main dish
b)   a pudding
c)   a starter.[9]

The Spanish test does not include queries into the candidates' motives.[10] Each administration produces a free booklet providing information for these tests. These exams might appear depersonalised, however the French one, and not the Spanish, is followed by an interview covering much the same terrain, but more extensively, and face to face. Moreover, as one British woman revealed to Fiona, she had to respond to cultural questions as a French woman would be expected to respond rather than expressing her own opinions. According to advisory websites for French citizenship, the length of the interview varies from a few minutes to a few hours. Some candidates state 'They were put through the mill. . . . Much may depend on the official in front of you and whether they are having a good or bad day'. One immigration advice consultant forewarned applicants that, for the interview, they might need 'A bit more knowledge than what is contained in the booklet, and if it comes across that you are going through the process just to get papers then they are going to spot it'.[11] Academic researchers underline the discretionary power of bureaucrats throughout these procedures, while both advisory websites and personal testimonies emphasise the process is 'often disheartening', 'a bit of an ordeal'.[12] This is intentional: as the French prime minister stated in 2019, 'Becoming French is demanding'.[13] Becoming Spanish is the diametrical opposite; it is recognised as the easiest EU nationality to gain (González Enríquez 2014). The speed of naturalisation in Spain is also one of the faster in the EU, that of France among the very slowest (González Enríquez 2014:5; Dag Tjaden 2018).

In France, which allows dual citizenship, a significant number of Britons have applied for naturalisation: they are ready to bear two passports, perhaps also to claim two nationalities (Table 6.1). Though gaining Spanish citizenship is easier and quicker, dramatically fewer Britons in Spain have done so compared with their counterparts in France. In 2018, 3,268 Britons acquired French citizenship, compared to 56 gaining Spanish citizenship (Eurostat).

Even though the numbers applying in each country have multiplied since the referendum, the number of British people seeking French citizenship is still over 58 times the number requesting Spanish citizenship. The difference between the national figures is all the more striking, given that the British population in Spain is very likely over double that in France.

At the Elche and Perigueux meetings held in 2017, the audiences were asked to raise their hands if they were considering, respectively, Spanish or French citizenship. An overwhelming majority of the French audience raised their hands, but few in the Spanish audience did. One activist campaigner opined that pragmatic considerations were the main reason for this reluctance: for instance, they worried about future provision of their pensions, or possible return to the UK to care, say, for elderly kin. If that were a key factor, it should of course apply equally to Britons in France. Some of the responses from the audience in Spain revealed a strong emotional attachment to their British citizenship which they were, at that stage, not

Table 6.1 The number of French and Spanish citizenships acquired by Britons 2008–18

Country of Citizenship	2008	2009	2010	2011	2012	2013	2014	2015	2016	2017	2018
France	230	231	205	261	335	354	279	374	517	1733	3268
Spain	48	23	56	49	21	50	67	28	44	54	56

Source: Eurostat https://appsso.eurostat.ec.europa.eu/nui/submitViewTableAction.do

willing to consider giving up. Their reaction exposes, in a way that our French examples do not, the importance of British nationality to some of these migrants. As one who was changing citizenship expressed, 'Maybe some will say I'm a traitor, but I feel like my country has betrayed me' (Buck 2017). According to the Chair of Bremain in Spain, British applicants for Spanish citizenship 'tend to be people who have been here for a very long time, and who have a special connection to the country. Many will be married to Spaniards' (Buck 2017).

On this reading, though Spanish citizenship is relatively easy to obtain, its acquisition is not a quick solution for those keen to stay, but a radical decision for those so committed to permanent residence in the country that they are prepared to formally denounce their British citizenship to the Spanish authorities, and risk the pragmatic benefits of doing so. Despite the fact that several hundreds of thousands of Britons live in the country, it seems, so far, that only a few hundreds have been disposed to affirm a new nationality, especially if it might come at the cost of displacing the one they were born and brought up with. By the same token, retaining British citizenship illegally but being treated by local bureaucrats as a Spanish citizen when they are in Spain appears an equally unacceptable step for many Britons residing there.

The evolution of citizenships: practised pathways or Isinic enactments?

In previous chapters we have analysed our material in either neo-Bourdieuesque or Isinic terms. In this chapter, the twin actions of individual migrants to secure residency or to gain citizenship appear cases where the use of exploiting one approach over the other is not, at first sight, quite so clear-cut.

Residency first: migrants registering in Spain or pursuing residency in France follow pathways scripted by their respective governments. This appears simple practice theory: they reproduce categorical boundaries laid down by overseeing administrations. This is clear for the Spanish case. However, in France, migrants who wish to remain appear to be acting in a manner edging towards the Isinic. True, they are behaving in a way generated by the state, but this is one largely unknown in practice, and they are applying in numbers surprising and sometimes overwhelming to bureaucrats who are, moreover, often reluctant to process their paperwork. French authorities' questioning of Britons' right to apply for residency permits highlights the ambiguity of these migrants' subject positions and the pressing novelty of their cumulative petitions. But the late-2018 response of French officialdom – to reinforce acceptable pathways, albeit as a temporary measure not a definitive end-point – reincorporates migrant action within bureaucratic structures. To go down this path, applicants have to fit themselves into the discrete categories of acceptable migrants: worker, student, retired, inactive. This is more social reproduction than social emergence. Since the referendum, academics, policy think tanks and the campaign groups, among others, have highlighted the diversity of Britons residing in France (and elsewhere on the Continent); they have called for a new system of registration to acknowledge this diversity (e.g. Benton et al. 2018). So far, the system has not been altered: Britons who wish to stay must adapt to the norms. A seeming phase of Isinic creativity has passed.

Citizenship second: UK migrants to the Tarn are becoming Irish or French citizens as well as British ones, in increasing numbers. They have won the right to bear two passports. This style of naturalisation is nothing new. To that extent it could be considered as but another constituent of people's broad habitus, albeit one seldom employed in the past. It thus fits without problem, as a pre-existing, formerly marginal procedure, into contemporary practice theory. An Isinic vocabulary does not appear particularly applicable or its use revealing here. In these spheres of action, migrants are not creatively pushing at categorical boundaries nor acting in hitherto-unknown ways at different scales in a manner surprising to national administrations or a supranational body, such as the EU. Instead they are taking advantage of civic pathways already mapped out by governments. Dual nationality is a long-established status, though its incidence has been historically low, geographically patchy, and socially uneven. But as a recent survey demonstrates (Harpaz and Mateos 2019), international circumstance is making this once unusual practice more common.

An Isinian might counter-argue that migrants who gain Spanish citizenship yet say nothing to their British consulate are acting in an innovative, but informal and covert manner. However, as several have pointed out, the relevant authorities are aware of this practice; it is just they decide to ignore or silently condone it. From what we can glean, this does not appear a new mode of operation. We might call this a space of unstated encounter where administrations implicitly connive with duplicitous migrants, to the benefit of all. British migrants with Spanish citizenship can remain in the country until death, yet may return anytime to their birthplace, if needs must; they are not upset by surrendering that symbolically charged object: a UK passport, now re-issued in its traditional navy blue. At the same time, the Spanish Government gains greater control over its population, increases its citizenry, and augments its tax revenue.

What does appear new is the public touting by national administrations of citizenship status to the rich. This move, by the governments of Malta and Cyprus, troubles the European Commission as it threatens to open a legal entrance into the EU for those with money whose origin is not always clear. This is an option few can exploit: Jeremy did not learn of any examples in his field-area; Fiona met one pair who had done so, and only heard of one other. This means to naturalisation is different from those earlier discussed for more than monetary reasons. Britons who acquire Irish citizenship on the grounds of parentage often admit the mild but affective dimension of this change; they are pleased, in however slight a manner, to honour the memory of their recent forebears. Similarly, in websites by, on, and for British migrants, interviews with the recently naturalised expose the sentimental statement made by acquiring citizenship: some claim it is a public expression of their long-term love for France, to which they are now openly committing themselves. In stark contrast, the purchase of citizenship by outsiders for nakedly instrumental reasons is often regarded askance (Shachar 2017). For this is to gain a valued status, of pragmatic benefit, moral depth, and emotional charge, not for fundamental reasons of birth, residence, or parentage, but thanks to a simple financial transaction. The underlying rationale to this ethical unease is that, if citizenship is so cherished, why can it be bought, as though off the shelf, by someone whose only connection to the country is commercial? This is not Isinic change; in neo-Bourdieuesque terms, it is an imposed innovation from above which opens its field to new players. In the process it stretches the idea of citizenship, into openly marketable realms, where morality now abuts money-making, and its customary link to territoriality is weakened.

The substance of this chapter also demonstrates the evolving content of forms of citizenship, at supranational, national, and personal levels, in ways which may overlap or conflict with the fought-for changes discussed in the previous chapter. Here we have seen citizenship, customarily tied to birth-state, now turned by some from a legal status into a legal commodity, for unfettered sale to high bidders. At the same time, naturalisation becomes more commonplace, and individual possession of multiple citizenships inches towards the norm. Yet for those with more than purely commercial, instrumental ties to citizenship, the idea continues to retain a malleable degree of affective force, ethical compass, and identifying orientation.

Notes

1 '?Cual es la diferencia entre los terminos 'ciudadanía' y 'nacionalidad?', Ministerio de Asuntos Exteriores, Gobierno de Espana. Available at www.exteriores.gob.es/ Consulados/ROSARIO/es/ServiciosConsulares/Paginas/Preguntas%20frecuentes/ preguntas%20registro%20civil/Adquisici%C3%B3n%20nacimiento%20%20 nacionalidad/Cu%C3%A1l-es-la-diferencia.aspx (Accessed 20 April 20). See also Luis Gálvez Muñoz, 2003, 'Sinopsis artículo 11', *Constitución Española*. Available

at https://web.archive.org/web/20050507165848/www.congreso.es/constitucion/constitucion/indice/sinopsis/sinopsis.jsp?art=11&tipo=2 (accessed 20 April 20).

2 As stated in the EU's Freedom of Movement Directives (see Directive2004/38/EC).

3 Due to be launched in the latter months of 2020.

4 One can also apply for citizenship though marriage to a French citizen.

5 Unknown to this individual, French language exceptions apply for people aged 60 and over, as well as for those with a disability or a chronic health condition.

6 This additional example remains unverified.

7 Benson and O'Reilly, who worked France and Spain respectively, though in different field-areas to us, highlight in their work this uncertainty and anxiety caused by Brexit (Benson 2020; O'Reilly 2020).

8 From October 2020, British citizens living in France will need to apply for 'withdrawal agreement' residence permits, which be obligatory for continued residence from 01 July 2021.

9 https://examenes.cervantes.es/sites/default/files/Modelo-CCSE-2018.pdf (accessed 14 May 2020).

10 For information on the French test, see file:///Users/p0066680/Downloads/Livret-du-citoyen_pageapage_5mars2015%20(1)%20(1).pdf; https://unetunisienneaparis.com/2018/01/01/99-questions-preparer-entretien-naturalisation/; on the Spanish one, see https://examenes.cervantes.es/es/ccse/examen (all accessed 14 May 2020).

11 The quote is from www.thelocal.fr/20191218/do-you-know-france-well-enough-to become french. See also www.superprof.us/blog/french citizenship test/; www.connexionfrance.com/Community/Your-News/What-really-happens-in-a-citizenship-application (all accessed 14 May 2020).

12 Fassin and Mazouz (2009:41), Bertossi (2010:17–18). Online examples are both the websites in n.vi, www.france24.com/en/20160629-how-become-french-citizen-british-brexit-france-naturalisation-process and www.completefrance.com/living-in-france/real-life/my-experience-of-applying-for-french-citizenship-1-4885398 (accessed 14 May 2020).

13 Eduoard Philippe, quoted in www.rfi.fr/en/france/20190506-tougher-language-tests-could-lengthen-migrants-wait-french-citizenship (accessed 14 May 2020).

7 Could there be a conclusion?

What have we demonstrated? That political agency among migrants is not confined to the underprivileged or multiply disadvantaged. Even incomers who are comfortably off have their own political subjectivities and will enter the political fray if they feel their fundamental values challenged or key interests threatened. The councillors stood to fight gross moral inequity. The campaigners organised to combat what they saw as the stupidity of Brexit. Some migrants, keen to remain without worry, learnt both to navigate and to contest the policy and practice of national bureaucracies.

What does surprise us is that we have needed to demonstrate this point. For very understandable and often laudatory reasons, the majority of migration researchers have focussed on the downtrodden, the imperilled, and the discriminated against. They hope, against hope, that their words and studies may contribute to amelioratory measures. Scholars of lifestyle migration have less interventionist agendas. They research a recently emergent phenomenon: the movement of the relatively well-to-do, who migrate not for the sake of security or jobs but to improve their already pleasurable existence, what they term 'quality of life'. Academics of underprivileged migrants tend to concentrate on the political structures and policies which keep them in their lowly place. Researchers of lifestyle migration, by definition, investigate the maximising strategies of lotus-eaters and other would-be hedonists, whether employed or not. Members of either academic camp have little to do with those in the other. It is another, unfortunate example of silo culture, where social scientists, for disciplinary reasons, keep to their sub-fields and tend to ignore the work of those intellectually nearby. In contrast, we have striven to avoid falling down any one silo but have attempted to meet on an interdisciplinary terrain, where concepts such as citizenship, belonging, political agency, morality, and habitus intermesh in productive ways. Whether or not we have succeeded, you can judge.

We wished to be as theoretically coherent as possible. For us, deployment of contemporary practice theory worked well as an overall frame for our analyses. It helped order the various strategies by different participants to make the most of, and develop their diverse forms of capital: economic, social, symbolic. But habitus, whether classical or neo-Bourdieuesque, is only somewhat adaptable, not endlessly flexible. When we wished to scrutinise the labour of the campaign groups in the fullness of their radically open-ended, self-constructed, co-evolving, context-creating process, habitus was not up to the diagnostic task. Here we turned to Isin's approach to enacting citizenship. Though he formulated his mode of analysis to follow the protests of the usually marginalised, and lacking citizenship status, we found it also fitted precisely the sort of innovative, multiscalar activities we were examining. The campaign groups' behaviours cut new paths, which in turn defined them.

Contemporary practice theory and Isin's approach are not incompatible opposites, as we showed in the preceding chapters. We do not see them as intellectual rivals for the same analytical space. They are of different scope, with distinct aims. Practice theory is more concerned with how societies continue, Isin's approach with how they can change. In O'Reilly's vision, practice theory is a very broad framing device (O'Reilly 2012). We can call it a productive means of organising large data in socially relevant ways. Isin's aims are far more focussed. He does not deal with whole societies but with activist sub-groups within them. His approach is not necessarily focussed on more localised processes as they may cut across a pre-existing socioscape of fields and hierarchies. It is not parasitic but rather revisionary of the already extant. To put that another way, if we saw his style as nestling within that of practice theory, it would be as though a cuckoo was in that nest.

Isin's approach can be re-considered in at least three aspects. First, Isin and his acolytes tend to study long-term, concerted campaigns. Yet, as demonstrated in the previous chapter when we discussed migrants' strategies to cut their own road through the labyrinth of French bureaucracy, one can identify liminal phenomena, where agents' enactment is observable but much less organised and somewhat fleeting. We might call this a phase of Isinic creativity. Second, Isin wants to emphasise the fluid nature of the dynamic processes he wishes to see studied. He also desires to escape the disabling dimensions of former theories by creating a novel analytical vocabulary. Yet his choice of terms is too static, emphasising 'acts' rather than process, so threatening to unravel a central thread of his approach. One could counterargue this is a necessary consequence of linguistic structure: every sentence has to have a noun. But isolating concentrated though brief stretches of processes as 'acts' is always open to charges of arbitrary delimitation and threatens to cramp the processual dimensions Isin is so keen to develop. Perhaps we should concentrate on processes of enacting rather than moments of 'acts'. In our narrative, no single 'event' or 'act' of the campaigners had a discrete, demonstrable effect: what was important was the rolling series of actions at different levels or scales whose processual effect was usually cumulative.

Third, Isin's approach is insufficiently self-reflexive. The activists he studies participate in a developing context, of which they and we their scholars are a part. Yet the present style of Isin's theory ignores the integral role of a researcher as a participant in the process he/she is studying: for instance, Fiona's own negotiations to secure residency in France are informed by and simultaneously inform those participants she speaks with. In the context of Brexit, the researcher's role may edge towards the central both because of the creative, open-ended nature of activism here, and because of internet activity. Two examples: first, at a Brexit seminar held at Southampton University in March 2017, a doctoral student studying one activist campaign group discussed his rapid incorporation into its inner circle.[1] When asked if he was now more an activist than an analyst, he replied, 'I analyse data, I make comments on positions taken, I do not suggest strategy'. Whether he could retain that distinction as the process developed remained open. Second, after Jeremy gave lunch at an Oxford college to a leader of another group, and invited other representatives to our 2017 Elche gathering, accounts of both meetings soon went up on their websites, as they did on our project webpages. We had hoped our webpages would become an integratory site of their own, bringing together otherwise disparate groups. In fact, the meeting itself, more than the website, fulfilled that function. In this sense, activist and analyst both play parallel but interactive games for similar justificatory ends, though of course the effect of each is wildly different. Either way, the point remains: we should not prematurely exclude ourselves from the exploratory, unfolding processes we study, even if our own role may be minor.

Over the course of our chapters we have repeatedly attempted to tease apart a tangle of ideas – citizenship, residency, status, home, and morality – used in novel ways and unforeseen connections. The British migrants who, at the Perigueux meeting, cried, 'But we are home!', wished to separate natal citizenship from place of residence yet retain some link between the two. Perhaps a more exact formulation is that they wanted official recognition that though their citizenship remained British, their home was mobile; in shifting residence successfully, they had created a new sense of home, which for some inched towards *heimat*, i.e. a strong sense of territorially grounded, lasting community. Maybe it is for this reason so many are prepared to consider a second citizenship, that of their new homeland, France.

The EU caused a more radical change. By creating a supranational citizenship, it broke the direct link between citizenship and nationality, and so allowed members of the EU to participate equally and fully as political agents within the municipal life of other EU countries. In the process, electoral candidates and elected councillors flesh out an emerging, novel EU form of municipal citizenship, one independent of a person's country of birth. Its importance, as we show, is not one of contribution to national politics but to creation of a liveable and just community which gives value to how the migrants think of themselves and where they live. At the same time, members of EU states have been able to exert their political agency and give substance to this new, supranational citizenship by fomenting citizens' initiatives: if seven members of different states can organise an e-petition signed by over a million EU-members, then the European Commission is bound legally to consider it, though it retains the power to reject. Launched in 2012, this measure, 'the first instrument of transnational participatory democracy known', is meant to put EU citizens on the same footing as the European Parliament and the European Council (Peñarrubia Bañón 2016:144). By the end of 2019, five initiatives had successfully reached the European Commission. However, the bureaucracy involved is burdensome, and their legislative consequence marginal, so far (Csehi and Puetter 2016:122; Longo 2019). These petitions are less Isinic enactments than enablements of European-stimulated legislation. The idea of an EU-citizenship has won some substance, but a functioning European polity of participatory citizens remains far off.

The governments of Malta and Cyprus have introduced a further twist: the open hawking of citizenship to rich outsiders. In this nakedly commercial manoeuvre, money rubs against morality and the customary link between citizenship and territoriality is weakened yet further. Increased naturalisations might make the holding of multiple citizenships ever more quotidian, but the unashamed instrumental exploitation of citizenship continues to cause unease (Shachar 2017): putting a price tag on this privilege weakens the affective charge, moral bearing, and identificatory orientation citizenship is still, for many, meant to provide. Their instrumental logic could be compared to the strategic intentions of Britons seeking French and Spanish citizenship. But as we have seen, the French Government is keen to corral these self-seeking claimants by making acquisition onerous. It does not want its citizenship devalued; its emotional force weakened. In this complex field, where citizenship intermeshes with affect, ethics, and identity, governments clash on how its status is to be understood while its EU version is dependent indirectly on popular agitation for its developing substance. We cannot formulate any conclusions here, just an interested tracing of this conceptual cartography and its continuing border changes.

We have also highlighted Hispano-French differences in bureaucratic policy and practice. Those differences continue. In France, Britons could not participate in the municipal elections of March 2020. Two months earlier, one migrant had written to Fiona, 'People have received letters from their mairie saying that we are no longer on the electoral list, councillors have had to stand down and can't stand for re-election'. In contrast, in March 2019, the Spanish Government agreed with HMG to retain the electoral powers registered British migrants already held. According to the Agreement, this 'will allow greater integration [of UK nationals] and will ensure their political and social rights'.[2] In France, the days of British councillors fomenting municipal change have ended. In Spain, they continue. Perhaps Britons in France will find or create new ways of integrating themselves into local society, developing their habitus in the process.

The rationale of the campaign groups has not remained static but evolved, while some migrants shifted position, especially with the Conservative electoral victory of December 2019, and the formal departure of the UK from the EU the following month. Some groups had already left the coalition (Bremain in Spain, RIFT); others disbanded (Europats, Fair Deal for Expats); a new one formed (France Rights). Many groups, forced to accept the reality of Brexit, have shifted focus: to secure as much as possible their members' valued rights, among them Votes for Life in the UK, for Britons living on the Continent. Moreover, worry about securing extant rights transcended Remain/Leave distinctions, for Brexit affects both groups equally. In France in late 2019 and 2020, Fiona found some Leavers sharing the same concerns as Remainers. One Leaver, not regretting their vote, told her, 'I never imagined needing residency permits, all this paperwork, driving licences'.

Campaigners had initially striven against their involuntary re-classification by HMG. Later, they became just as agitated about being ignored. In autumn 2019, Bremain in Spain complained, 'We feel really forgotten', and that the Spanish Government was doing more for Britons there than its British counterpart. By February 2020, faced with a new, strong Conservative government, Bremain highlighted key issues at that time as its invisibility and its stereotyping by Tory politicians and the British media.[3] In political terms, they had been marginalised and felt they were now either neglected or caricatured. In the first years of their campaign, activists could see themselves not as bit-part actors but as creative, influential protagonists, indirectly empowered by HMG, fighting to mould the continuing process of which they were a part. Those heady days appear to be over.

Throughout our text, we have striven to expose the constitutive role of the affective. Until relatively recently, too many ethnographies of political movements tended to marginalise this dimension in their accounts. Too often the result is an overly sociologistic account which at its worst comes across as alienating and unconvincing. In opposition to these a-emotional portraits, we believe we have demonstrated the integral, motivating role of the affective within the processes we analyse. The anti-Brexit protestors, at their most successful, created an affective political economy, bound together in an empowering state of indignation. We recognise the emotional responses of individual migrants, whether joyous or devastated by the Leave result. But we here stress the broader power of channelling common feeling for political end. The activists' efforts display how effective a motor the organisation of righteous anger can be. Without acknowledgement of the affective and its pervasive contribution to migrants' political subjectivities, our text would have been thin, grossly lopsided, and misleading.

But does the example of their activity remain? The earlier battle against expropriation in the Valencian community suggests it will not. That campaign had a successful outcome, but memory of their activity has faded quickly . In a zone of high mobility, collective memory can be shallow. To analyse further developments or suggestive permutations in the nexus of concepts we have considered, chances are one will have to look elsewhere.

Notes

1 'The legacy of Brexit: mobility and citizenship in times of uncertainty', day-long seminar, ESRC Centre for Population Change, University of Southampton, 31 March 2017: attended by Jeremy.
2 As defined in the UK/Spain Agreement on the participation in certain elections of nationals of each country resident in the territory of the other. March 2019. Available at: www.gov.uk/government/publications/cs-spain-no22019-ukspain-agreement-on-the-participation-in-certain-elections-of-nationals-of-each-country-resident-in-the-territory-of-the-other (accessed 10 May 2020).
3 www.bremaininspain.com/articles/brexit-brit-expats-in-malaga-stage-huge-protest-saying-uk-has-forgotten-us/;
www.bremaininspain.com/news/grassroots-for-europe-conference-25-jan-2020/ (accessed 17 May 2020).

References

Abascal, M. and Baldassarri, D. 2015. Love they neighbour? Ethnoracial diversity and trust reexamined, *American Journal of Sociology*, 121:722–82

Ackers, L. and Dwyer, P. 2004. Fixed laws, fluid lives: the citizenship status of post-retirement migrants in the European Union, *Ageing and Society*, 24:41–75

Aledo, A., Jacobsen, J. and Selstad, L. 2012. Building tourism in Costa Blanca: second homes, second chances?, in Nogués-Pedregal, A. (ed.), *Culture and society in tourism contexts*. Bingley: Emerald, 111–39

Arrighi, J.-T. 2014. *Access to electoral rights: France*. Report RSCAS/EUDO-CIT-ER 2014/1. Florence, Italy: EUDO Citizenship Observatory

Barbero, I. 2012. Expanding acts of citizenship. The struggles of Sinpapeles migrants, *Social and Legal Studies*, 21:529–47

Barou, J. and Prado, P. 1995. *Les Anglais dans nos campagne*. Paris: L'Harmattan

Beigel, F. 2009. 'Sur les Tabous Intellectuels': Bourdieu and academic dependence, *Sociologica*, 2–3. Available at www.rivisteweb.it/doi/10.2383/31370 (accessed 31 May 2017)

Benson, M. 2011. *The British in rural France: lifestyle migration and the search for a better way of life*. Manchester: University of Manchester Press

———. 2013. Living the "real" dream in *la France* profonde? Lifestyle migration, social distinction, and the authenticities of everyday life, *Anthropological Quarterly*, 86:501–26

———. 2019. Brexit and the classes politics of belonging: the British in France and European belongings, *Sociology*, 54:501–17, https://doi.org/10.1177/0038038519 885300

———. 2020. *Brexit and the British in France*. Project Report. London: Goldsmiths. Available at: https://doi.org/10.25602/GOLD.00028222 (accessed 20 June 2020)

Benson, M. and Lewis, C. 2019. Brexit, British people of colour in the EU-27 and everyday racism in Britain and Europe, *Ethnic and Racial Studies*, 42:2211–28

Benson, M. and O'Reilly, K. 2009. Migration and the search for a better way of life: a critical exploration of lifestyle migration, *The Sociological Review*, 57: 608–25

Benson, M. and Osbaldiston, N. (eds) 2014. *Understanding lifestyle migration: theoretical approaches to migration and the quest for a better way of life*. London: Palgrave Macmillan

Benton, M., Ahad, A., Benson, M., Collins, K., McCarthy, H. and O'Reilly, K. 2018. *Next steps: implementing a Brexit deal for UK citizens living in the EU-27*. Brussels: Migration Policy Institute Europe

Bermúdez, A. and Escrivá, A. 2016. La participación política de los inmigrantes en España: elecciones, representación y otros espacios, *Anuario de CIDOB de la Inmigración 2015–16*. www.cidob.org. Available at: https://www.cidob.org/es/articulos/anuario_cidob_de_la_inmigracion/2015_2016/la_participacion_politica_de_los_inmigrantes_en_espana_elecciones_representacion_y_otros_espacios (accessed 12 April 2020)

Bertossi, C. 2010. *Country report: France*. Florence, Italy: EUDO Citizenship Observatory. Available at https://cadmus.eui.eu/bitstream/handle/1814/19613/France.pdf?sequence=1&isAllowed=y (accessed 14 May 2020)

Birkvad, S.R. 2019. Immigrant meanings of citizenship: mobility, stability, and recognition, *Citizenship Studies*, 23:798–814

Boon, J.A. 1983. *Other tribes, other scribes: symbolic anthropology in the comparative study of cultures, histories, religions, and texts*. Cambridge: Cambridge University Press

Bossuet, L. 2006. Peri-rural populations in search of territory, *Sociologia Ruralis*, 46:214–28

Bourdieu, P. 1977. *Outline of a theory of practice*. Cambridge: Cambridge University Press

———. 1990. *The logic of practice*. Stanford, CA: Stanford University Press

———. 2005. Habitus, in Hillier, J. and Rookby E. (eds), *Habitus: a sense of place*, 2nd ed. Aldershot: Ashgate, 43–9

Bousiou, P, 2008. *The nomads of Mykonos: performing liminalities in a 'queer' space*. Oxford: Berghahn

Brändle, V.K. 2018. Reality bites: EU mobiles' experiences of citizenship on the local level. *Journal of Ethnic and Migration Studies*, 46:2800–17, https://doi.org/10.1080/1369183X.2018.1524750

Bremain in Spain. 2020. Dual citizenship – can you be a Spanish citizen and a British one?, *Bremain in Spain*, 1 ii. Available at www.bremaininspain.com/news/dual-citizenship-can-you-be-a-spanish-citizen-and-a-british-one/ (accessed 20 April 2020)

Buck, T. 2017. Spain's UK expats face Brexit citizenship dilemma, *Financial Times*, 17 March

Buller, H. 2008. Du côté de chez Smith: reflections on an enduring research object, in CERAMAC (ed.), *Les Étrangers dans les Campagnes. Actes du colloque franco-britannique de géographie rurale. Vichy 18 et 19 mai, 2006*. Clermont-Ferrand: Presses Universitaires Blaise Pascal

Buller, H. and Hoggart, K. 1994. *International counterurbanisation: British migrants in rural France*. Aldershot: Avebury

Burchianti, F. and Zapata-Barrero, R. 2017. From participation to confinement: challenges for immigrants' incorporation in political parties in Spain, *Ethnic and Racial Studies*, 40:830–50

Butler, J. 2004. *Undoing gender*. London: Routledge

Cannadine, D. 2013. *The undivided past: humanity beyond our differences*. New York: Alfred A. Knopf

Cantat, C., Sevinin, E., Maczynska, E. and Birey, T. (eds) 2019. *Challenging the political across boundaries: migrants' and solidarity struggles.* Budapest: Central European University

Carr, R. 1986. Introduction, in Macaulay, R. (ed.), *Fabled shore: from the Pyrenees to Portugal.* Oxford: Oxford University Press, 1–3

Centre for Ageing Better. 2020. *Doddery but dear? Examining age-related stereotypes.* Available at www.ageing-better.org.uk/sites/default/files/2020-03/Doddery-but-dear. pdf (accessed 6 April 2020)

Clout, H. 2006. Rural France in the new millennium: change and challenge. *Geography*, 91:205–17

Collard, S. 2010. French Municipal Democracy: cradle of European citizenship? *Journal of Contemporary European Studies*, 18:91–116

———. 2013. Evaluating European citizenship through participation of non-national European citizens in local elections: case studies of France and the UK, *Cuadernos Europeos de Deusto*, 48:135–73

Collins, K. and O'Reilly, K. 2018. *What does freedom of movement mean to British citizens living in the EU27? Freedom, mobility, and the experience of loss.* Available at https://ukandeu.ac.uk/wp-content/uploads/2018/11/Freedom-of-Movement-report.pdf (accessed 8 April 2020)

Collinson, P. and Kollewe, J. 2019. British and want an EU passport? Here's where you can apply, *The Guardian*, 7 September

Cook, J., Long, N.J. and Moore, H. 2016. Introduction. When democracy "goes wrong", in Cook, J., Long, N. and Moore M. (eds), *The state we're in: reflecting on democracy's troubles.* New York: Berghahn, 1–26

Csehi, F. and Puetter, E. 2016. The European Union's political citizens: rights, practices, challenges, and alternative models of participation, in Seubert, S. and van Warden F. (eds), *Being a citizen in Europe: insights and lessons from the open conference, Zagreb 2015.* Available at www.uu.nl/sites/default/files/being-a-citizen-in-europe-insights-and-lessons-from-the-open-conference-zagreb-2015.pdf (accessed 25 June 20), 121–22

Dag Tjaden, J. 2018. *Access to citizenship and its impact on immigrant integration: handbook for France.* Florence, Italy: Migration Policy Group, European University Institute

Daloz, J.-P. 2013. *Rethinking social distinction.* London: Palgrave Macmillan

Diez Medrano, J. 2008. Europeanization and the emergence of a European Society. *IBEI Working Papers.* Available at http://papers.ssrn.com/sol3/papers.cfm?abstract_id=1086084 (accessed 04 April 2020)

Directive 2004/38/EC of the European Parliament and of the Council, 29 April 2004: on the right of citizens of the Union and their family members to move and reside freely within the territory of the Member States: amending 290 Regulation (EEC) No 1612/68 and repealing Directives 64/221/EEC, 68/360/EEC, 72/194/EEC, 73/148/EEC, 75/34/EEC, 75/35/EEC, 90/364/EEC, 90/365/EEC and 93/96/EEC. *Official Journal of the European Union*, L158

Drake, H. and Collard, S. 2008. A case study of intra-EU migration: 20 years of "Brits" in the Pays d'Auge, Normandy, France, *French Politics*, 6:214–33

Dugot, P., Laborderie, S. and Taulelle, F. 2008. *Midi-Pyrénées: region d'Europe.* Midi-Pyrénées: CRDP

Edwards, J. n.d. *Fault lines: Europe, Brexit and anthropology.* Douglass Distinguished Lecture, Society for the Anthropology of Europe, given Vancouver November 2019

Engel, M. 2016. "I don't want to go back with nothing": the Brexit threat to Spain's little Britain, *The Guardian*, 27 May

Eurostat. 2020. Available at https://appsso.eurostat.ec.europa.eu/nui/submitViewTable Action.do (accessed 29 March 2020)

Favell, A. 2008. *Eurostars and Eurocities: free movement and mobility in an integrating Europe.* Oxford: Blackwell

Fassin, D. and Mazouz, S. 2009. What is it to become French? Naturalization as a Republican rite of institution, *Revue française de sociologie*, 50:37–64

Favell, A. and Recchi, E. 2009. Pioneers of European integration: an introduction, in Recchi, E. and Favell, A. (eds), *Pioneers of European identity: citizenship and mobility in the EU*. Cheltenham: Edward Elgar, 1–25

Ferbrache, F. 2011a. British immigrants in France: issues and debates in a broadening research field, *Geography Compass: Environment and Society*, 5:737–49

———. 2011b. *Transnational spaces within the European Union: the geographies of British migrants in France.* University of Plymouth. Unpublished Ph D thesis

———. 2019a. Acts of European citizenship: how Britons resident in France have been negotiating post-Brexit futures, *Geography*, 104:81–8

———. 2019b. Local electoral rights for non-French residents? A case-study analysis of British candidates and councillors in French municipal elections, *Citizenship Studies*, 23:502–20

Ferbrache, F. and Yarwood, R. 2015. Britons abroad or European citizens? The negotiation of (trans)national space and citizenship by British migrants in France, *Geoforum*, 62:73–83

Gemperle, M.I. 2009. The double character of the German "Bourdieu". On the twofold use of Pierre Bourdieu's work in the German-speaking social sciences, *Sociologica*, 1. Available at www.rivisteweb.it/doi/10.2383/31370 (accessed 31 May 2017)

Golding, J. and Morgan, J. 2017. *UK citizens in Europe: Towards an alternative White Paper on the European Union (Notification of Withdrawal) Bill.* Available at https://britishineurope.org/wp-content/uploads/2017/02/alternative_white_paper_presented_by_uk_citizens_in_europe.pdf (accessed 19 November 2018)

González Enríquez, C. 2014. El precio de la ciudadanía española y europea, *Real Instituto Elcano. Estudios internacionales y estratégicos.* Available at www.realinstituto elcano.org/wps/portal/rielcano_es/contenido?WCM_GLOBAL_CONTEXT=/ elcano/elcano_es/zonas_es/demografia+y+poblacion/ari22-2013-gonzalez-enriquez-precio-ciudadania-espanola-europea (accessed 14 May 2020)

Guild, E. 2016. The legal framework: who is entitled to move?, in Bigo, D. and Guild, E. (eds), *Controlling frontiers.* Routledge: Abingdon, 14–48

Hare, A.J.C. 1873. *Wanderings in Spain.* London: Strahan

Harpaz, Y. and Mateos, P. 2019. Strategic citizenship: negotiating membership in the age of dual nationality, *Journal of Ethnic and Migration Studies*, 45:843–57

Heiberg, M. 1989. *The making of the Basque nation*. Cambridge: Cambridge University Press

Ho, E. 2008. Citizenship, migration and transnationalism: a review and critical interventions, *Geography Compass*, 2:1286–300

Hoggart, K. and Buller, H. 1994. Property agents as gatekeepers in British house purchases in rural France, *Geoforum*, 25:173–87

———. 1995. British home owners and housing change in rural France, *Housing Studies*, 10:179–98

Huber, A. and O'Reilly, K. 2004. The construction of *Heimat* under conditions of individualised modernity: Swiss and British eldery migrants in Spain, *Ageing and Society*, 24:327–51

Huete, R. and Mantecón, A. 2012a. La participación política de los residents británicos y alemanes en España: el caso de San Miguel de Salinas, Alicante, *Revista de geografía Norte Grande*, 51:81–93. Available at www.redalyc.org/articulo. oa?id=30023283009 (accessed 30 May 2017)

———. 2012b. Residential tourism or lifestyle migration: social problems linked to the non-definition of the situation, in Moufakkir, O. and Burns P. (eds), *Controversies in tourism*. Wallingford: CAB International, 160–73

Huete, R., Mantecón, A. and Estévez, J. 2013. Challenges in lifestyle migration research: reflections and findings about the Spanish crisis, *Mobilities*, 8:331–47

INE (Instituto Nacional Estadística). n.d.a. *Población extranjera por nacionalidad, comunidades, sexo, y año*. Available at www.ine.es/jaxi/Tabla.htm?path=/t20/e245/p08/l0/&file=02005.px&L=0 (accessed 12 April 2020)

———. n.d.b. *Estadística del Padrón Continuo a 1 de enero de 2001. Datos a nivel nacional, comunidad autónoma y provincia*. Available at www.ine.es/jaxi/tabla. do?path=/t20/e245/p04/a2001/l0/&file=00000010.px&type=pcaxis (accessed 11 April 2020)

INSEE. 2010. *Population*. Chiffre clés: Évolution et structure de la population. Available at www.insee.fr/fr/themes/theme.asp?theme=2&sous_theme=0&type=2&nivgeo=7&submit=Ok. (accessed 10 May 2020)

Isin, E. 2008. Theorizing acts of citizenship, in Isin, E. and Nielsen, G.M. (eds), *Acts of citizenship*. Chicago: University of Chicago Press, 15–43

———. 2009. Citizenship in flux: the figure of the activist citizen, *Subjectivity*, 29:367–88

———. 2012. *Citizens without frontiers*. London: Bloomsbury

———. 2013. Claiming European citizenship, in Isin, E. and Saward, M. (eds), *Enacting European citizenship*. Oxford: Oxford University Press, 19–46

———. 2017. Performative Citizenship, in Shachar, A., Bauböck, R., Bloemraad, I. and Vink, M. (eds), *The Oxford handbook of citizenship*. Oxford: Oxford University Press, 500–23

Isin, E. and Saward, M. 2013. Questions of European citizenship, in Isin, E. and Saward, M. (eds), *Enacting European citizenship*. Oxford: Oxford University Press, 1–18

Isin, E. and Turner, B. 2002. Citizenship studies: an introduction, in Isin, E. and Turner, B. (eds), *Handbook of citizenship studies*. London: Sage, 1–10

Janoschka, M. 2009. *Konstruktion Europaischer Identitaten in Raumlich-Politischen Konflikten*. Weisbaden: Franz Steiner Verlag

———. 2010. Prácticas de ciudadanía europea. El uso estratégico de las identidades en la participación política de los inmigrantes comunitarios, *Arbor*, CLXXXVI(744):705–19

———. 2011. Habitus and radical reflexivity: a conceptual approach to study political articulations of lifestyle- and tourism-related mobilities, *Journal of Policy Research in Tourism, Leisure & Events*, 3:224–36

Jenkins, R. 1992. *Pierre Bourdieu*. London: Routledge

Joppke, C. 2019. The instrumental turn of citizenship, *Journal of Ethnic and Migration Studies*, 45:858–78

Kallio, K.P., Wood, B.E. and Häkli, J. 2020. Lived citizenship: conceptualising an emerging field, *Citizenship Studies*, https://doi.org/10.1080/13621025.2020.1739227

Kemp, C. 2010. Building bridges between structure and agency: exploring the theoretical potential for a synthesis between habitus and reflexivity, *Essex Graduate Journal of Sociology*, 10:149–57

Kofman, E., Phizacklea, A., Raghuram, P. and Sales, R. 2000. *Gender and international migration in Europe: employment, welfare, and politics*. London: Routledge

Lagarde, V. and Di Pietro, V. 2019. Brexit before Brexit. Consequences of Brexit's anticipations on the British entrepreneurs in France between 2016 and 2019, in Sacco, M. (ed.), *Brexit a way forward*. Malaga: Vernon Press, 29–88

La Vanguardia. 2018. *Diecisiete municipios españoles tienen mayoría de población extranjera*, 25 April. Available at www.lavanguardia.com/vida/20180425/44299 9739349/diecisiete-municipios-espanoles-tienen-mayoria-de-poblacion-extranjera. html (accessed 12 April 20)

Lawson, M. 2016. *Identity, ideology and positioning in discourses of lifestyle migration: the British in the Ariège*. London: Palgrave

Lazar, S. (ed.) 2013. *The anthropology of citizenship: a reader*. Oxford: Wiley-Blackwell

Lister, M. 2008. Europeanization and migration: challenging the values of citizenship in Europe?, *Citizenship Studies*, 12:527–32

The Local. 2020. *Thousands sign petition after Briton in Dordogne refused French citizenship*. Available at www.thelocal.fr/20200202/thousands-sign-petition-after-briton-in-dordogne-refused-french-citizenship (accessed 25 March 2020)

Longo, E. 2019. The European Citizens' initiative: too much democracy for EU polity?, *German Law Journal*, 20:181–200

López-Sala, A. 2019. 'You're not getting rid of us'. Performing acts of citizenship in times of emigration, *Citizenship Studies*, 23:97–114

Lori, N.A. 2017. Statelessness, "in-between" statuses, and precarious citizenship, in Shachar, A., Bauböck, R., Bloemraad, I. and Vink, M. (eds), *The Oxford handbook of citizenship*. Oxford: Oxford University Press

Macaulay, R. 1949. *Fabled shore: from the Pyrenees to Portugal*. London: Hamish Hamilton

Macbeth, A. 2019. Brits in Spain hope for dual citizenship legislation in 2019, *The Local*, 8 January. Available at www.thelocal.es/20190108/brits-in-spain-hope-government-could-open-up-to-dual-citizenship-legislation-in-2019 (accessed 20 April 2020)

MacClancy, J. 2000a. *The decline of Carlism.* Basque Book Series, Centre for Basque Studies, Reno. Reno, NV: University of Nevada Press

———. 2000b. The predictable failure of a European identity, in Axford, B., Berghahn, D. and Hewlett, N. (eds), *Unity and diversity in the new Europe.* Oxford: Peter Lang, 111–28

———. 2007. *Expressing identities in the Basque arena.* Oxford: James Currey

———. 2015. Fear and loving in the West of Ireland: the blows of County Clare, in MacClancy, J. (ed.), *Alternative countrysides: anthropological approaches to rural Western Europe today.* Manchester: University of Manchester Press, 143–68

———. 2016. Down with identity! Long live humanity!, in Eriksen, T.H. and Schober, E. (eds), *Identity destabilised: living in an overheated world.* London: Pluto, 20–41

———. 2019. Before and beyond Brexit: political dimensions of UK lifestyle migration, *Journal of the Royal Anthropological Institute,* 25:368–89

Margetts, H. 2013. The internet and democracy, in Dutton, W.H. (ed.), *The Oxford handbook of internet studies.* Oxford: Oxford University Press, 421–40

McNevin, A. 2006. Political belonging in a neoliberal era: The struggle of the Sans-Papiers, *Citizenship Studies,* 10:135–51

———. 2012. Undocumented citizens. Shifting grounds of citizenship in Los Angeles, in Nyers, P. and Rygiei, K. (eds), *Citizenship, migrant activism and the politics of movement.* London: Routledge, 165–83

Méndez-Lago, M. 2010. La participación de los extranjeros comunitarios en las elecciones municipales en España: 1999, 2003 y 2007', in Moya Malapeira, D. and Viñas Ferrer, A. (eds), *Sufragio y participación política de los extranjeros extracomunitarios en Europa.* Barcelona: Fundació Pi Y Sunyet, 503–30

Menjivar, C. 2006. Liminal legality: Salvadoran and Guatemalan immigrants' lives in the United States, *American Journal of Sociology,* 111:999–1037

Mindus, P. 2017. *European citizenship after Brexit: freedom of movement and rights of residence.* London: Palgrave

Ministère de l'intérieur. 2018. Enquiry on British Participation Levels in 2014 Municipal Elections. [letter] (Personal communication, 17 July 2018)

Ministry of Interior. 2018. *Grant of the Cypriot citizenship to non – Cypriot entrepreneurs/investors through the "Scheme for Naturalization of Investors in Cyprus by exception.* Ministry of Interior, Republic of Cyprus. Available at www.moi.gov. cy/moi/moi.nsf/All/36DB428D50A58C00C2257C1B00218CAB (accessed 21 January 2019)

Morrison, D. and Compagnon, A. 2010. *The death of French culture.* Cambridge: Polity

Mouzelis, N. 2007. Habitus and reflexivity; restructuring Bourdieu's theory of practice, *Sociological Research Online,* 12. Available at www.socresonline.org.uk/12/6/9. html (accessed 22 March 2018)

Oliver, C. and O'Reilly, K. 2010. A Bourdieusian analysis of class and migration, *Sociology,* 44:9–66

Oliveri, F. 2012. Migrants as activist citizens in Italy: understanding the new cycle of struggles, *Citizenship Studies,* 16:793–806

ONS (Office of National Statistics). 2018. *Living abroad: British residents living in the EU: April 2018*. Available at www.ons.gov.uk/peoplepopulationandcommunity/populationandmigration/internationalmigration/articles/livingabroad/april2018 (accessed 25 March 2020)

O'Reilly, K. 2000. *The British on the Costa del Sol. Transnational identities and local communities*. London: Routledge

————. 2007. Intra-European migration and the mobility-enclosure dialectic, *Sociology*, 41:277–93

————. 2012. *International migration and social theory*. Basingstoke: Palgrave Macmillan

————. 2017. The British on the Costa del Sol twenty years on: a story of liquids and sediments, *Nordic Journal of Migration Research*, 7:139–47

————. 2020. *Brexit and the British in Spain*. Project Report. London: Goldsmiths. Available at https://doi.org/10.25602/GOLD.00028223 (accessed 20 June 2020)

Ortiz, A. 2015. Los extranjeros, infrarrepresentados en los municipios donde son mayoría, *El Mundo*, 3 iv. Available at www.lavanguardia.com/politica/20150503/54431002135/los-extranjeros-infrarrepresentados-en-los-municipios-donde-son-mayoria.html (accessed 12 April 2020)

O'Toole, F. 2019. Dreams of empire, blitz spirit, a country in decline . . . how competing visions of the past are driving Brexit, *The Guardian, Review supplement*, 2 November, 12–13

Paz, A.I. 2019. Communicating citizenship, *Annual Review of Anthropology*, 77:48–93

Peñarrubia Bañón, M. 2016. The European Citizens' Initiative as an instrument of participatory democracy. New formula to fill the democratic deficit of the European Union?, in Seubert, S. and van Warden F. (eds), *Being a citizen in Europe: insights and lessons from the open conference, Zagreb 2015*. Available at www.uu.nl/sites/default/files/being-a-citizen-in-europe-insights-and-lessons-from-the-open-conference-zagreb-2015.pdf (accessed 25 June 20), 136–45

Peró, D. 2014. Recognising migrants' practices of citizenship and their impact, in *Migration: the COMPAS anthology*. Oxford: COMPAS. Available at https://compasanthology.co.uk/recognizing-migrants-practices-citizenship-impact/ (accessed 23 April 2020)

Postill, J. 2010. Introduction: theorising media and practice, in Bräuchler, B. and Postill, J. (eds), *Theorising media and practice*. Oxford: Berghahn, 1–26

————. 2012. Digital politics and political engagement, in Horst, H.A. and Miller, D. (eds), *Digital anthropology*. London: Bloomsbury, 165–84

Preuss, U.K., Everson, M., Koenig-Archibugi, M. and Lefebvre, E.L. 2003. Traditions of Citizenship in the European Union. *Citizenship Studies*, 7:3–14

Putnam, R. 2007. *E pluribus unum*: diversity and community in the twenty-first century. The 2006 Johan Skytte Prize Lecture, *Scandinavian Political Studies*, 30:137–74

Puzzo, C. 2007. British migration to the Midi-Pyrenees, in Geoffrey, C. and Sibley, R. (eds), *Going abroad: travel, tourism, and migration. Cross-cultural perspectives on mobility*. Newcastle: Cambridge Scholars Publishing, 110–18

Recchi, E. 2008. Cross-state mobility in the EU, *European Societies*, 10:197–224

Recchi, E. and Favell, A. (eds) 2009. *Pioneers of European identity: citizenship and mobility in the EU*. Cheltenham: Edward Elgar

Remarque Koutonin, M. 2015. Why are white people expats when the rest of us are immigrants?, *The Guardian*, 13 March. Available at www.theguardian.com/global-development-professionals-network/2015/mar/13/white-people-expats-immigrants-migration (accessed 30 May 2017)

RIFT. 2020. *RIFT sounds the alarm bell for Brexit British in France*. Report by RIFT. Available at www.remaininfrance.fr/rift-sounds-the-alarm-bell (accessed 28 April 2020)

Rodríguez, A. 2013. *Access to electoral rights: Spain. Report RSCAS/EUDO-CIT-ER 2013/15*. Florence, Italy: EUDO Citizenship Observatory

———. 2018 *Report on political participation of mobile EU-citizens: Spain. RSCAS/GLOBALCIT-PP 2018/22*. Florence, Italy: European University Institute

Rogaly, B. 2020. *Stories from a migrant city: living and working together in the shadow of Brexit*. Manchester: Manchester University Press

Ronkainen, J.K. 2011. Mononationals, hyphenationals, and shadow-nationals: multiple citizenship as practice, *Citizenship Studies*, 15:247–63

Scott, S. 2004. Transnational exchanges amongst skilled British migrants in Paris, *Population, Space and Place*, 10:391–410

———. 2006. The social morphology of skilled migration: the case of the British middle class in Paris, *Journal of Ethnic and Migration Studies*, 32:1105–29

Service-Public.fr. 2020. *Titre de séjour d'un travailleur citoyen UE/EEE/Suisse*. Available at www.service-public.fr/particuliers/vosdroits/F16003 (accessed 25 March 2020)

Seubert, S. 2018. EU citizenship and the puzzle of a European *political* union, in Seubert, S., Eberl, O. and van Waarden, F. (eds), *Reconsidering EU citizenship: contradictions and constraints*. Cheltenham: Edward Elgar, 21–41

Shachar, A. 2017. Citizenship for sale? in Shachar, A. and Bauböck, R., Bloemraad, I. and Vink, M. (eds), *The Oxford handbook of citizenship*. Oxford: Oxford University Press, 789–816

Shore, C. 2000. *Building Europe: the cultural politics of European integration*. London: Routledge

Simó-Noguera, C.-X., Herzog, B., Torres, F., Jabbaz, M. and Giner Monfort, J. 2005. Asociacionismo y población extranjera en la Comunidad Valenciana, *Cuadernos electrónicos de filisofía de derecho*, 12. Available at https://dialnet.unirioja.es/servlet/articulo;jsessionid=B26A7E50087437DE91D6796088E34C01.dialnet02?codigo=1307311 (accessed 30 May 2017)

Sitwell, S. 1950. *Spain*. London: Batsford

Sorge, A. 2009. Hospitality, friendship and the outsider in Highland Sardinia, *Journal of the Society for the Anthropology of Europe*, 9:4–12

Staeheli, L.A. 2008. Citizenship and the problem of community, *Political Geography*, 27:5–21

Staeheli, L.A., Ehrkamp, P., Leitner, H. and Nagel, C.R. 2012. Dreaming the ordinary: daily life and the complex geographies of citizenship, *Progress in Human Geography*, 36:628–44

Swartz, D. 2013. *Symbolic power, politics and intellectuals: the political sociology of Pierre Bourdieu*. Chicago: University of Chicago Press

Teindas, N. 2009. *Les Britanniques dans le Sud-Ouest: greffe ou rejet?*, Paris: Editions Praelego

Theodossopoulos, D. 2013. Infuriated with the infuriated? Blaming tactics and discontent about the Greek financial crisis', *Current Anthropology*, 54:200–9

Thorold, P. 2008. *The British in France: visitors and residents since the Revolution*. London: Continuum

Thorpe, A. 2018. *Notes from the Cévennes: half a lifetime in Provincial France*. London: Bloomsbury

Tomé da Mata, E. 2015. Participación de los ciudadanos de la Unión Europea en las elecciones al Parlamento Europeo y elecciones locales en España, *RIPS. Revista de investigaciones políticas y sociológicas*, 14:27–62

Tremlett, G. and Chislett, W. n.d. Dual nationality for Brits who have resided in Spain for more than 10 years, *Change.org*. Available at www.change.org/p/dual-nationality-for-brits-who-have-resided-in-spain-for-more-than-10-years (accessed 20 April 2020)

United for All Ages. 2020. *Together in the 2020s: twenty ideas for creating a Britain for all ages by 2030*. Available at https://efeea61d-ae40-4f75-bfce-8a7be79f7237. fllesusl.com/ugd/98d289_3f3291f2d4094c2793a3aef8ffaae58a.pdf (accessed 6 April 2020)

van Houtum, H. and Boedeltje, F. 2009. Europe's shame: death and the borders of the EU, *Antipode*, 41:226–30

Wilson, S. 2019. Bremain in Spain's sue Wilson: how Brexit changed our lives, *Dispatches Europe*, 22 May. Available at https://dispatcheseurope.com/bremain-in-spains-sue-wilson-how-brexit-changed-our-lives/?fbclid=IwAR0H1kHrGgIhQo-H0VVozQxeS3LAUQIapcoHh6DeaRZy8M5QEne44iBrA9w (accessed 17 April 2020)

Wise, M. 2015. *The Tarn* [email] (Personal communication, 18 July 2015)

Wise, M. and Gibb, R. 1993. *Single market to social Europe: the European Community in the 1990s*. Harlow: Longman Group

Wood, P.M. and Gilmartin, M. 2018. Irish enough: changing narratives of citizenship and national identity in the context of Brexit, *Space and Polity*, 22:224–37

Woodward, I. and Emmison, M. 2009. The intellectual reception of Bourdieu in Australian social sciences and humanities, *Sociologica*, 2–3. Available at www.rivisteweb.it/doi/10.2383/31370 (accessed 31 May 2017)

Yarwood, R. 2014. *Citizenship*. Abingdon: Routledge

Index